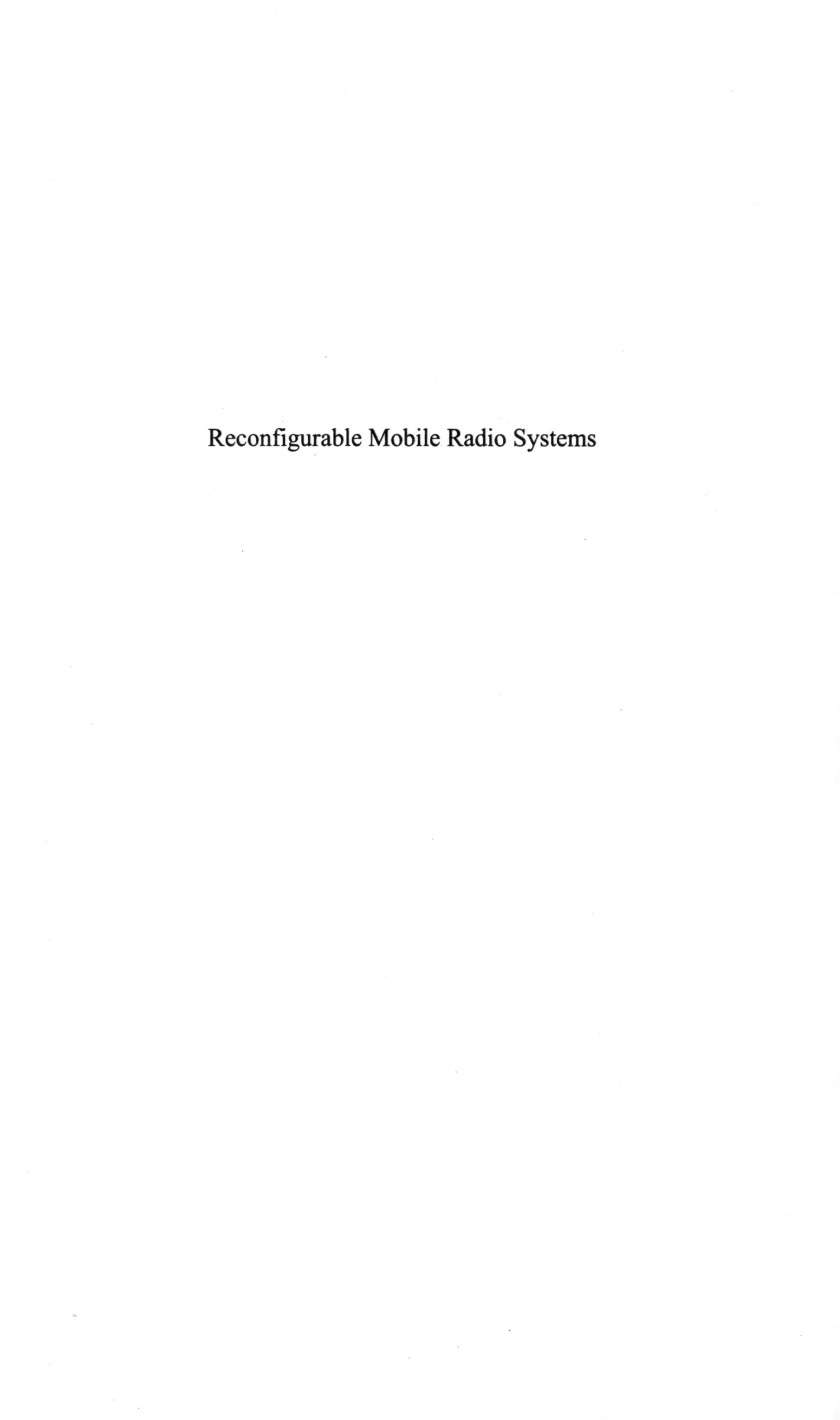

Reconfigurable Mobile Radio Systems

Reconfigurable Mobile Radio Systems

A Snapshot of Key Aspects Related to Reconfigurability in Wireless Systems

Edited by
Guillaume Vivier

First published in France by Hermes Science/Lavoisier entitled "Les systémes radiomobiles reconfigurables"

First Published in Great Britain and the United States in 2007 by ISTE Ltd.

First South Asian Edition 2007

Apart from any fair dealing for the purposes of research or private study, or criticism or review, as permitted under the Copyright, Designs and Patents Act 1988, this publication may only be reproduced, stored or transmitted, in any form or by any means, with the prior permission in writing of the publishers, or in the case of reprographic reproduction in accordance with the terms and licences issued by the CLA. Enquiries concerning reproduction outside these terms should be sent to the publishers at the undermentioned address:

ISTE Ltd.
6 Fitzory Square
London WJT 5DX
UK

ISTE USA
4308 Patrice Road
Newport Beach, CA 92663
USA

www.iste.co.uk

© ISTE Ltd., 2007
© LAVOISIER

The rights of Guillaume Vivier to be identified as the author of this work has been asserted by him in accordance with the Copyright, Designs and Patents Act 1988.

This edition is for sale in the Indian subcontinent only. Not for export elsewhere.

ISBN 10: 1 905209 46 0
ISBN 13: 978 1 905209 46 0

Library of Congress Cataloging-in-Publication Data

[Systemes radiomobiles reconfigurables, English] reconfigurable mobile radio systems: reconfigurable mobile radio systems:
a snapshot of key aspects related to reconfigurability in wireless systems/edited by Guillaume Vivier.
p.cm.
Originally published: France: Les systèmes radiomobiles reconfigurables.
Hermes Science/Lavoisier,
Includes index.
ISBN 10: 1 905209 46 0
ISBN 13: 978 1 905209 46 0
1. Mobile communicaton systems. 2. Wireless communication systems.
I. Vivier, Guillaume.
TK6570.M6S97 2007
621.382--dc22

2007001890

British Library Cataloguing-in-Publication Data
A CIP record of this book is available from the British Library
ISBN 978 1 905209 46 0

Printed and bound in India by Brijbasi Art Press Limited, I-72, Sector-9, Noida, U.P., India.

Table of Contents

Introduction

Digital mobile wireless communication and the Internet have undergone a fantastic growth in the last few years and, despite originating from two different worlds, they are converging. This convergence corresponds to the evolution of mobile systems towards the highest broadband data transmissions (GSM, EDGE/GPRS, UMTS then HSDPA), while the computing world gets equipped with wireless technologies such as Wi-Fi or Wi-Max.

Due to progress and the integration of digital technologies, communication systems (either network systems or user terminals) have become multimode, i.e. able to operate in multiple standards within various frequency bands. Applications, protocol stacks or transceivers are adaptive, optimizing processing based on environmental conditions, users' preferences and provisioned services. Soon, all equipment and systems will be entirely reconfigurable, meeting the vision of software radio developed initially in the military field.

Obviously, this increase of flexibility within the equipment in return requires a more complex management: how should this reconfiguration capability be managed? Why, when and how should reconfiguration be used? Is it possible to design a radio system with a flexible spectrum allocation? What would be the roles of service providers, operators, regulators and users? Who controls what? Which are the technological limits of the reconfiguration? Will the services and users know how to make adjustments?

In this book, our objective is to put forward answers to these questions, by gathering around the theme of reconfigurable radiomobile systems a panel of experts, coming from various horizons: universities, manufacturers and operators, all of them being time users.

Chapter 1, written by Guillaume Dorbes, addresses the new models of services and the applications induced by the technological evolutions associated with the convergence of the fixed and mobile communication systems, showing increasing reconfigurability capability, and bearing in mind that more and more people use them.

The implementation of reconfigurable mobile radio systems requires more and more software. Chapter 2, written by Antoine Delautre and Yann Denef introduces the concept of object modeling in the software context and presents its interest in the reconfigurable systems. A review of the most relevant developments and standards, especially in the military field, is then proposed.

Chapter 3, written by a team of France Telecom R&D engineers, looks into the design of a software radio terminal. The interests of such an approach as well as its underlying constraints are described through concrete examples, at the level of the hardware, of the software and also at the terminal overall architecture.

An example of concrete design of a reconfigurable terminal and a base station is then described in Chapter 4. The mobile communications team group in the Eurecom institute presents a UMTS-TDD software-defined radio platform whose protocol stacks are entirely implemented in C under Linux and thus are totally reconfigurable on request.

This full software approach is then completed by Chapter 5, which discusses hardware reconfiguration. Iannis Krikidis, Lirida Naviner and Jean-Luc Danger propose improvement axes of a *Rake* receiver, by automatic reconfiguration. Various reconfiguration strategies as well as the associated performances are also presented.

Sébastien Roy and Jean-Yves Chouinard extend the topic of hardware reconfiguration by covering in Chapter 6 the specific problem of the multiantenna transmissions (MIMO). Indeed, MIMO is considered by all the recent mobile radio systems. After a presentation of the MIMO systems and their performances, this chapter discusses their practical implementation in reconfigurable architectures.

Chapter 7 focuses on a key element of the software-defined radio: analog-to-digital converters. Firstly, Patrick Loumeau, Lirida Naviner and Jean-François Naviner describe the role of the converters in typical receivers as well as the current performances. They then present various converter structures, especially the most promising ones to be used in reconfigurable mobile radio systems.

Finally, to conclude this book, David Grandblaise extends the flexible spectrum management concept to the most general cognitive radio topic. In fact, the final stage of the reconfigurable mobile radio systems would be to be able to use any standard in any band. Chapter 8 thus presents the (r)evolutions of thought around the spectrum and its management at the regulatory level as well as on concrete techniques which enable to share it efficiently.

This book thus provides an extended overview of the most relevant subjects related to reconfigurable mobile radio systems. A reader involved in this field can stop to the most technical chapters and go into more detail by means of the numerous references given at the end of each chapter, while a reader in a hurry may satisfy himself picking up the sufficient information by simply going through the chapters, in order to understand the problems related to the reconfigurability of a wireless communication system.

Chapter 1

Services and Adaptive Uses

1.1. New networks and new uses

1.1.1. *Broadband mobile radio systems: why do it?*

Why would we cover adaptive services and uses within a work dedicated to reconfigurable mobile radio systems? We can raise this question because the two issues are very distant: one is very technological and focused on wireless networks, whereas the other, equally (but only) technological, is also focused on the person who will use these networks.

Why should wireless broadband networks be deployed? For what kind of services? How can these networks improve our everyday life? What is going to change due to these networks? Will we have access to new services or will the existing services be adapted to this mobile dimension?

This part of the study tries to respond to these questions and perhaps raises others. We will try to understand the uses leading to the use of these networks, having taken time to think about the problem in order to analyze what our "customers" expect from these networks and then we will think of the technological and human factors which will condition the large scale adoption of these mobile networks.

Chapter written by Guillaume DORBES.

At first there was the fixed network, then came the mobile one, but is it not the same story?

1.1.2. *From Internet services on a voice network to voice services on an Internet network*

Will mobile Internet be the future of fixed Internet? We can believe this, since the progress of technologies and behaviors have increased so much in the last 10 years. Therefore, even if this work deals with mobile networks, the aim of this chapter is to present "services and adaptive uses" by addressing the fixed telephony networks which, in this regard, are rich in considerations regarding the new mobile services.

Let us cast our minds back ... In that period, in France there was only a single fixed telephony operator that used to sell us minutes of communication. Then we witnessed the appearance of modems, those boxes that made the Internet connection possible by dialing a phone number billed at the price of a local communication. The telephony services were kings and the transport of Internet services using these wires dedicated to spoken communication was tolerated.

Since then, things have evolved. For around €30 per month, a home can benefit from unlimited national phone communications, faster Internet access and a range of television channels.

What happened in the meantime? A technology of broadband access network has completely changed the way services were provided: ADSL (*asymmetric digital subscriber line*[1]). Before, the Internet was a minority service of a residential telephony network: now residential telephony has become a minority service (in volume of octets) of an Internet access network.

This trend has been started by operators new to the marketplace who, wanting to find their place in the sun, did not have any other choice but surfing among the possibilities offered by the new broadband network technologies. The success of ADSL is very easy to explain: more and faster services for less money.

1 Technology enabling a broadband data communication using a traditional phone line (copper wire).

In the light of this report, there is of course the massive adoption of IP[2] protocol of Internet communication for the quasi-totality of exchanges. This major technological evolution has also led to a new chance for communication uses and actors.

1.1.3. *From telephony to interpersonal communication*

The adoption of Internet technology by the majority of people has also led to the adoption of new communication services that were inherent to it. Email is probably the most significant example: reserved for a group of initiated people 10 years ago, it is now at the heart of communication for most companies and is one of the professional services with the highest bandwidth use. Other new usages of the Internet such as instant messages have also been massively adopted.

It must be mentioned that the leading actors of these new means of communication are rarely telephony operators or access networks. Thus, the two greatest global email operators are mainly Internet services providers:

– number 1: Yahoo (213 million users at the end of 2002[3]);

– number 2: Microsoft Hotmail (145 millions users in January 2004[4]).

This new situation thus enables these actors to take a position in the traditional telephony market as one of the Internet service providers. Here and now, this position is obvious in a country such as Japan[5] where an Internet operator has more than 10% of the market share in the residential telephony market. This very strong penetration of Internet services into communication services tends to erase the barriers that existed between historical spoken services on the one hand and data transport services on the other hand.

Once more, the consequence is that the balance tilts toward Internet services due to the new possibilities they offer such as videoconferencing, exchange of multimedia folders, presence functions, etc.

2 Internet Protocol: http://www.commentcamarche.net/internet/ip.php3.
3 http://docs.yahoo.com/info/investor/metrics.html.
4 http://www.microsoft.com/presspass/newsroom/msn/factsheet/hotmail.asp.
5 http://www.journaldunet.com/0401/040128voipjapon.shtml.

1.1.4. *From charged to free: the value evolution*

With the Internet we get accustomed to new commercial practices. Many people would not imagine for a moment to be obliged to pay in order to receive an email address. This has become a free service like many other Internet services.

This gratuity obviously has a price! This might be the obligation to check your messages on a web page with commercials, filling in of a form with information on your profile or through additional paid services such as SMS notification or an anti-virus. The basic communication services are always free. Nothing stops you anymore from staying all day long "chained" to your PC chatting with your cousins from far away.

This situation is for the moment the exclusive benefit of the fixed Internet where only the access is billed with a fix amount depending on the desired traffic. Since services are independent from access – including the communication services – if they are billed, this is not necessarily done anymore by the access operator. This is nevertheless a burdening tendency because the public begins to get used to this type of practice considered as an acquired advantage. One of the issues that we shall debate again further concerns the validity of this approach in the mobile world and the economic models which will find an echo among the users.

1.1.5. *From the end-to-end controlled session to the best effort culture*

I will not discuss here again the IP and the basic principles of Internet protocols that I consider known, but one of the reasons for Internet adoption: simplicity. In fact, if the Internet technology has known such an adoption, it is mainly because it offers an environment having maximum possibilities – e.g. a worldwide network open for everybody – with minimum constraints on this network: little intelligence and if a piece of information is lost on the way, then it is retransmitted within a reasonable period of time.

Could this "do the best" principle involve quality problems? The Internet detractors think so and they do not hesitate to say that the current VoIP[6] services do not always meet the expectations. Sometimes they are right, but those who remember having tried to call from their car on a winter's night in 1988, when all the Parisian area was blocked due to the snow, would say that even for the non-

6 VoIP: *Voice-over-IP*.

Internet mobile telephony, when resources are not sufficient, there is no miracle solution. Therefore, is the famous QoS[7], lost with the IP, a legend without a real foundation? Polemics aside, we can at least ask ourselves this question, especially if this potential disadvantage is compensated by a significant decrease of costs.

1.1.6. *The new services of the new networks*

Thus, a possible summary of the generalized adoption of dial-up Internet and then of the broadband Internet has five points:

– "phagocytosis" of copper wires dedicated to the telephony by broadband Internet access;

– massive adoption of the web as support for Internet content services;

– globalization of communication services, either vocal or not, simultaneous (e.g. conversations) or asynchronous (e.g. messages);

– free of charge communication services and evolution of the value chain;

– adoption of IP as basic protocol for the entire traffic.

Why are these heavy trends forming the fixed Internet not extended also to the mobile world? Are we that different as nomad communicators from sedentary connected users? What does it mean then to propose services adapted to this mobile use? Before answering these questions, let us take into consideration what guides our choices and our usage as customers of these technologies and services.

1.2. Mobile communications customers

1.2.1. *Mobile service user: a communicating customer*

The human being has actually changed very little in the last few thousands years. His basic needs are always the same: to drink, to eat, to sleep, to reproduce. In order to satisfy these vital needs, the human being has to communicate and be informed. All that we are searching for, as human beings, in the world of NTIC[8] can be reduced to this simple statement. Services with a universal success, such as fixed or mobile telephony, the web, search engines or email are a part of these

7 QoS: *Quality of Service*.

8 New information and communication technologies.

preoccupations. Of course, there are many other popular services but they do not have such a universal character.

It can therefore be considered that at best, these needs are not going to be fundamentally modified by the new generations of broadband radio networks. On the other hand, besides these "basic" needs, the consumer society in which we live got us used to satisfying our human weaknesses. Hence, the marketing gurus are trying to satisfy our taste for power, our impatience, our laziness, our greed or our ego. Applying these ideas to the field of communication and information services comes to satisfy the basic needs of the communicating customer: more, faster, simpler, less expensive communication services.

This is the operating mechanism of the consumer society and this explains very simply the success of the Internet or of the mobile telephony in this context.

1.2.2. *The successful teachings of mobile telephony and the Internet for the new generation services*

Let us first define a new generation mobile service. To simplify, let us simply say that it refers to a service available via wireless equipment connected to the Internet due to one or more wireless networks. This covers thus different target terminals such as the telephone, a PDA, a laptop or a car. Their common feature is the capability of being connected to the Internet generally continuously without a copper wire. The access means could be thus as various as a public or professional residential Wi-Fi network, a GPRS or UMTS network, WCDMA such as Wi-Max.

This diversity of equipment and networks makes very likely the fact that during a common working day we have to get connected to the Internet using different service providers. At home, we have this access provided by a residential ISP[9], on the way to the office we have the cellular telephone carrier which seems to be favored, at the office, the company ensures our communication logistics and finally, aviation and rail transport companies may provide us with this access service as part of the price of our tickets.

However, apart from these access services, who will provide our Internet communication services? One of these actors or another one? The problem is more intense because, as we have seen, there is a total independence between access

9 Internet Service Provider.

services and communication services. Moreover, the relative captivity of the fixed Internet access – because it is attached to a physical place – will disappear due to the mobile character of our daily life.

In this context, we can expect a struggle among all these actors for gaining the privilege to take care of our communication services in order to obtain profit. It can be observed here that we are very far from the economic model of traditional telephony. This situation will be reached even sooner if the equipment manufacturers of telecommunication and information technology will provide their third or fourth generation of wireless network technologies[10]; this is anyway in line with the history regarding the competitive pressure among these same equipment manufacturers.

Facing this profusion of complex technologies, the user will be a little lost. Its natural selection method will remain unchanged, a kind of "Darwinism[11]" of service consumption: more, faster, simpler and less expensive. Hence, it is up to the operators to progress in this context.

1.2.3. *The communicating customer and his values*

Due to all these evolutions, the communicating customer, who is the object of desire for the operators, is thus confronted with new preoccupations resulting from the emergence of these technologies. We list below a few examples.

1.2.3.1. *Compatibility with the present and its practices*

Even if a service – e.g. vocal communication – becomes free of charge, we expect that it will not question overnight a certain number of ingrained practices such as numbering and addressing plan.

1.2.3.2. *Membership and availability*

Today, we are accustomed to using phonebooks in order to search for people. In their large public form, we mostly find there fixed phone numbers but very few cellular numbers, incidentally the new VoIP numbers, very rarely email addresses and never the *nicknames*[12] of the online messenger. And yet, all these features are a

---

10 This terminology covers the UMTS, HSDPA or Wi-Max type of networks offering bandwidths from a few hundreds of kilobits/s to a few megabits/s.

11 Information on Darwin: http://www.infoscience.fr/histoire/portrait/darwin.html.

12 It is a method of identification used by online messengers, other discussion forums.

part of our personal or business cards. Furthermore, this situation is probably going to extend if one wishes to have references of other communication services such as web pages, webcams or multimedia data servers. The counterpart of this rich offer of communication means is finally nowadays the fragmentation of research means to contact a person.

1.2.3.3. *Cost optimization*

If we still consider that during a certain period of time the bandwidth in a wireless network is a rare value and therefore expensive, we will try to find an acceptable compromise between the service comfort and the cost. Hence, a mobile email system will be able to check the message arrival and transfer its content using any available network but it will wait to get connected to a broadband or free network in order to download the attachments.

1.2.3.4. *Security*

The almost daily actuality has to face the new danger represented by the viruses promptly created and widespread, damaging our communication tools, especially via email. It is the opportunity to criticize both the operating system providers and the communication applications such as Internet itself. What is going to happen when our entire communication – in all its forms – is based on the Internet? It is very possible that a customer will not accept that a virus or an evil-minded person steals his identity especially during a phone call or a discussion in a professional forum.

1.2.4. *Mobility based acceleration*

Our sensitivity of the communicating customer based on these values increases so much more when we become mobile communicating customers. In fact, mobility has a magnifying effect due to the usage factors that it introduces.

1.2.4.1. *Terminal size and its interaction modes*

By nature, the mobile equipment is smaller and more restrictive in terms of interface with its owner. The question here is not to pass from one application to another with a mouse click or to open a list containing several hundreds of items. Having access to a mobile service means thus necessarily having simpler and more intuitive acçess than, for example, on a PC.

1.2.4.2. *Multi-network environment*

By definition, mobile equipment is supposed to follow us all day long and through all our different means of multi-network access, i.e. multi-operators. Nevertheless, our single concern as a customer is to benefit from our favorite communication and information services as simply as possible. In order to use these services in a mobile environment, it is necessary to erase, or at least hide, the technical constraints resulting from this mobility.

And yet, the multi-network environment is a significant source of benefits. The first one is doubtlessly cost minimization: why in fact paying for high cost communications while we have a several megabits/s network either free of charge, or by subscription irrespective of the traffic supported. This is exactly what enables the existence of a residential or corporate WLAN gateway use based on a dual mode mobile equipment (i.e. with two network interfaces: GPRS/3G and WLAN). This same type of configuration equally makes it possible to benefit from the very local character of a WLAN access point in order to design services which use the attachment to such a network like a localization service. For example, it is possible to use the attachment information of a dual mode terminal with domestic or professional WLAN access point in order to infer that a user is either at home or in the office and to transfer the incoming fixed calls into a VoIP call on the same terminal. We can see from such an example that the benefit of such a procedure is double for the user: an inherent economic advantage of the end-to-end IP service on the one hand and the facility offered by the use of a single personal equipment for the whole of residential or mobile and personal or professional communications, on the other hand.

The counterpart of usage facilities pertains to the management of the potential instability of the links and *handover* mechanisms (transition from a radio cell to another). There are different options to manage this type of a problem: some of them use the technologies of the operator access network such as UMA[13] services, whereas others, called *loose coupling*, are based on independent customer/server procedures of access networks. Such approaches are not necessarily antagonistic but result from complementary uses depending especially on the actors involved in the application of such services.

13 UMA – *unlicensed mobile access*: http://www.umatechnology.org.

1.2.4.3. *Service heterogenity*

The diversity of networks, the heterogenity of terminal equipment, as well as the geographical mobility of people and their preferences are parameters that condition the communication strategies among fixed or mobile individuals. It is thus possible to favor the interactive video communication for certain people and to have a fallback strategy toward the messenger, the vocal or even textual communication according to the constraints of the networks, the availability or personal choices.

It is not only about taking into account the constraints of the access network or terminal, it is also about adjusting to the reality of a certain usage in a given instant. Therefore, depending on whether his vehicle is moving or stopped, a driver will be able to benefit from a vocal or graphic interface in order to check his different messages. Communicating in a mobile multimedia world means to be able to adjust to an inherently more dynamic context that has immediate consequences on its communication ways.

1.2.5. *Adaptability as a mobility value*

Hence, from the previous examples, we see that due to the burdening technological tendencies structuring the new communication tools and to their application/operation within a mobile context, the new mobile Internet services lead to constraints of adaptability of existing services in order to be integrated in the context of a real use mechanism that can be adopted by a mass market.

This adaptability will definitely not consist only of modifying the existing services regarding their object (e.g. a vocal communication is what it is) or their specific architecture (e.g. a message leaves or arrives from a server to a customer), but of creating a mediation intended to take charge of the mobility by what it implies in terms of new complexities and resources, heterogenity, or even discomfort.

Besides, while the members of 3GPP[14] are interested in new mobile Internet services and work for their specifications through the IMS[15], they also adopt this same approach intended to adjust what already exist to the mobile constraints:

> "In order to achieve access independence and to maintain a smooth interoperation with wireline terminals across the Internet, the IP multimedia subsystem attempts to be conformant to IETF 'Internet standards'. Therefore,

---

14 *3rd generation partnership project:* http://www.3gpp.org.

15 *IP multimedia subsystem.*

> the interfaces specified conform as far as possible to IETF "Internet standards" for the cases where an IETF[16] protocol has been selected, e.g. SIP[17]."

In the context of a "mobilized" Internet, in the IMS by the 3GPP, we will see some tendencies and important technological elements which must be taken into account in order to offer these adaptability services.

1.3. Technological and adaptability factors of mobile services

1.3.1. *A microcomputer inside each pocket*

The first important factor concerning the constraints and possibilities of mobile Internet refers to the terminal itself. In fact, since around 2003, the great majority of mobile phones on the market have a feature enabling their similarity to a pocket microcomputer irrespective of their functional perimeter: they have an open operating system.

Be it Symbian™[18], Windows Mobile™[19], PalmOS™[20] even Linux[21], they make it possible to run applications developed by anyone in the world.

Installing an application on a mobile phone is no longer the privilege of terminal manufacturers or of their operator customers. Therefore, it is not by chance that from now on many companies[22] have made the distribution of mobile equipment applications their main activity. In addition, these phones equally offer a Java™[23] execution environment which facilitates the development and deployment of these applications.

In this context, we see that it is possible for an independent service provider to develop applications which could offer IP communication functions following the example of what happened on the PCs. Even if the editors of operating systems have

16 Internet Engineering Task Force: http://www.ietf.org.
17 Extract taken from the introduction of the IMS specification document reference 3GPP TS 23.228 v6.6.0.
18 http://www.symbian.com.
19 http://www.microsoft.com/windowsmobile.
20 http://www.palmsource.com.
21 http://www.linuxdevices.com.
22 For example, Handango: http://www.handango.com.
23 J2ME: Java 2 Platform Micro-Edition http://java.sun.com/j2me.

a special place in this landscape, the horizon remains open. In fact, in the PC/fixed Internet world, a certain number of actors have started their activity in a start-up mode before appearing as leading actors in their field. It is, for example, the case of ICQ™[24] with more than 250 million downloads and 150 million registered users. More recently, it is the telephony software Internet Skype™[25] which, in a few months, had more than 15 million downloads.

With a pocket microcomputer and an open operating system, we thus enter the universe where everything is possible in terms of mobile Internet services, including that of making phone calls without a telephonic network but only with an Internet access network. In fact, from the operating system, its IP protocol stacks and the multimedia interfaces of the terminal (e.g. MIDP[26]) we can then develop all kinds of applications which meet the customers' demands and expectations as described in the previous section.

1.3.2. *An Internet or a juxtaposition of intranets?*

If one day you have configured your cell phone in order to have access to WAP or MMS services, you will have discovered that the IP addresses of the gateways have sometimes the form of 192.168.X.Y. You may have also used this same address "192.168.X.Y" with the same values X and Y when you have configured one piece of equipment of your domestic network connected to your ADSL modem. What could seem surprising is in fact a normal thing because your mobile operator considers its access network as an intranet, equal to your domestic network or that of your company.

If you are lucky enough to already have a multi-network mobile terminal, which operates for example in GPRS/EDGE and in Wi-Fi at the same time, you will have to manage these hops from intranet to intranet in order to access your mobile communication in optimal economic conditions. Moreover, if you have an ongoing communication during these hops, you will want that this should be clearly done by favoring, for example, the low cost and large bandwidth networks.

---

24 http://www.icq.com.

25 http://www.skype.com.

26 MIDP – *Mobile Information Device Profile:* http://java.sun.com/products/midp.

Based on these examples that we can live nowadays, we easily discover the dichotomy between the access issues and those of the service:

– the goal of the access is to have allocated a generally perennial IP address and to have offered a path to the Internet through a gateway whose address you do not know *a priori* as it is seen by the rest of the world;

– the goal of a service is to offer you communication means, to exchange information and to inform you independently of these addresses – non-universal – by making sure that you can be identified in a non-ambiguous way and potentially on different terminals.

This trend which started in the fixed Internet and which led to offer dynamic addressing services[27] is therefore even more relevant within a mobile context.

Even if we can think that the IPv6[28] could settle a certain number of problems related especially to the availability of addresses, it is not sure that, for simple security reasons and the reflexes of its reserved domain, the "public" or private intranets would be abandoned.

Thus, these customer/server mechanisms should be understood in the context of mobile services so that they could be operated within a mediation service which could, for example, include the selection of networks. In this type of service we find an extension of the service logic such as in the Mobile IP[29] protocol. The benefit of such an IP mobile approach is, among others, its total compatibility with the IMS specifications naturally completing the communication mechanisms such as those offered by SIP[30].

For the mobile user of services, all these technical considerations are at best abstruse and in the worst case they represent a real brake for the adoption of the aforesaid services. In fact, we touch here a global problem which consists of associating a real person to one or more addresses which are generally dynamic, inside one or more universal sets (telephonic numbering or Internet addressing), in order to contact him in multiple ways (call, message, data transfer, etc.) through one or more access networks.

---

27 Service consisting of offering a name of a stable Internet domain (e.g. www.toto.com) for services which are the object of an unstable IP address. Examples: http://www.dyndns.org, http://www.no-ip.com, etc.

28 IP version 6: http://www.commentcamarche.net/internet/ipv6.php3.

29 IETF RFC 3344: http://www.ietf.org/rfc/rfc3344.txt?number=3344.

30 Session Initiation Protocol: http://www.ietf.org/html.charters/sip-charter.html.

1.3.3. *On the convergence of universal sets or how to contact a person*

Within this new world which intends to crash the habits, the practices and the technologies of telephony and the Internet, we witness new preoccupations such as those consisting of being able to contact a person due to or in spite of all the communication means which he has. Actually, we are facing a paradox: the more communications means we have, the more attention we have to pay to the contact information we give to third parties. Thus, if a few years ago it was obvious that practically all the phone numbers were published in a phone book, today this is not necessarily the basic rule, especially if we speak about the number of a cell phone. All the same, it is not unusual to have two, even three email addresses generally distributed: that from the office, that for private correspondence and that we are obliged to have in order to use an Internet service but which we know is going to be "spammed"[31] all day long.

Meanwhile, abusive and dishonest commercial practices such as online attacks against our digital address books have made us be more prudent. It is normal that after having received many times at home a call from the salesman of "X" cuisine, we do not want to receive them anymore on our mobile at the office during meetings with our customers. Let us imagine what will happen tomorrow when our SIP phone address may look like "alice@wonderland.com". Actually, is it really a SIP address or is it just an email address except for a last name in a messenger program?

Will too many names kill the name? Besides the aphorism, we really feel that this multitude of communication means could complicate our life instead of simplifying it. This is an investigation field where a certain number of approaches is possible. Will solving the contact problem only be the concern of HSS[32] or are there possible alternatives and improvements? What about the yellow, white, pink pages, the search engines and corporate phone books? Who has never tried to take a taxi by using any means, getting off in an unknown place, the only concern being to arrive fast? Thus, even for a simple phone book service, the mobility leads to the introduction of new requirements such as those related to proximity.

---

31 Receipt of a large number of unsolicited messages.

32 HSS: *home subscriber server*. Set of basic functions of the 3rd generation mobile network data.

1.3.4. *Proximity as a way to address the mobile services*

"Tell me about myself. This is what I am interested in[33]."

What a suitable way to summarize our way to address the world, but it is not the only one. This is also a parameter which shapes our practices related to communication and information. It is about practices that many of us use in relation to stimulations of our communication environment. In fact, we can define this environment as a sequence of proximity circles around us which we generally define explicitly.

Firstly, there are the *people close to us*. They are those to whom we dedicate the majority of our communication. Family or friends, colleagues, they are an entire group of individuals with which we have quasi-daily or even multi-daily contacts. The relationships with the people close to us are characterized by a certain availability: we are ready to interrupt what we were doing in order to communicate with the people close to us and we probably check the messages from them before those from others. In Internet terminology, they are the members of the *buddy list* from our messenger.

As for *the acquaintances,* their number is much larger. They are those people whose contact information we already have because we have already used it once or because we think it could be useful. Our proximity is less strong than with the people close to us, but these are people about whom we know *a priori* the communication context, either professional or personal. Hence, we will answer a communication solicitation from an acquaintance. The list of these acquaintances forms our *address book*, either digital or in hard copy.

Besides these two circles, there is the *rest of the world*: people that we do not know and with whom we might communicate some day (with a small minority). Within this third circle of proximity, we will search for our contacts in public and professional directories, such as search engines.

This classification strongly conditions our communication means. Thus, we accept a lot of contact with the people close to us: an instant message with online answer, an SMS during a meeting, a multimedia message or a videoconference with his webcam. So many communication means which are only a little or not at all practical for acquaintances with whom we prefer – conventionally – the phone or

33 Extract taken from *Parlez-moi de moi*, a song by Guy Béart. See http://www.paroles.net.

email which are not so intrusive. By nature, communication with the *rest of the world* will probably disturb us if it is not our initiative.

This attitude with respect to our proximity circle is even more significant the more we are involved in the communication world. Thus, we understand better why video-telephony or videoconference have not yet known the success expected, all technical constraints removed. To send your image to a third party is mainly a situation that few of us appreciate especially if this third party is an unknown person. It is the same situation for the cell phone number which is mentioned by few of us in public or professional directories. This lack of motivation is related to the same phenomenon: we do not want to be disturbed in a personal manner by whomever, all the time, whenever – this being the exact nature of mobile telephony.

Therefore, for mobile services, all the dimensions of proximity have to be considered, be it affective, intellectual, professional, geographical or social proximity. Mainly, we communicate according to our interests and the technological tools that we have must take into account these dimensions in order to make them suitable for our personal expectations and to thus facilitate our life.

The first solution attempts in order to take into consideration these egocentric preoccupations already exist. It is for example the presence server, which provides a first level of information about the selected correspondents. This can also be geo-localized proximity services based on different technologies such as GPS or identification of network cells. But it is still a long way off to find a mobile multimedia communication service that fully satisfies our needs as mobile consumer in all its dimensions and in all its integrity.

1.3.5. *The jungle of networks or how can we communicate in a hostile environment?*

Our integrity as a communicating individual is actually menaced by the multiple attacks which could disturb our contacts. Thus, some time ago, during a forum for high technology, a kind sponsor offered free Wi-Fi for the area where hundreds of lecturers and participants were strolling with their laptops and other digital assistant devices. Thanks to this offer, I got connected to the Internet using my favorite network card in order to check my emails and other services. After a few minutes, the *firewall*[34] of my PC was overwhelmed by many alert messages as a result of many intrusion attempts. Did some of the participants at this forum want to do some "active industrial monitoring"? It always happens that as soon as we plug active computing equipment into an open environment such as the Internet or an uncontrolled intranet, we are the target of actions which are similar to hacking actions.

We can think that this is a problem of computing victims and it is right to think so. Nevertheless, we must not forget that the fixed and mobile Internet services are in fact operated in the world of information technologies (see section 3.1) and therefore they suffer identical effects.

Besides the recent waves of viral attacks by email, it has surely happened to you that you have received messages from people you have never heard of. These messages have been sent by viruses unbeknown to the listed transmitters. For the moment, the consequences are unpleasant but seldom serious.

On the other hand, we can imagine that the target of such an attack can be your new generation cell phone trying to answer to an Internet call from someone whose identity has been stolen by any virus or with dishonest intentions. The virus would then take another dimension by attacking more profoundly our personal environment through the phone.

Our relation with a phone is actually more involving, more permanent and more personal by its nature as well as more limited in its interactions than that with a computer, even a laptop. That is why we want to be able to rely on it in the same way and for all the usage we may need.

1.3.6. *How can we carry our home in our pocket?*

The secure environment of our home is a value that we try to find at any price wherever we are. It is first of all the automobile world that invites us to consider our car as an extension of our home. Some gastronomic critics sometimes praise one restaurant or another because they produce food "just like at home". Finally, the same thing applies to the NTICs where this value prevails in many places: *home page*, *home location register*[35], *home network*, etc.

---

34 "Firewall" is a system which makes it possible to protect a computer against the intrusions coming from the network http://aww.alcatel.com/group/cto/ri/menu/index.htm.
35 HLR: database of a mobile network.

These terms clearly translate our concern to identify a known and comforting model. The mobility has itself a value in total contradiction with these preoccupations because it is related to freedom. Free movement, removal of geographic constraints and "pin wire", choice of communication means. Thus, the stakes in new mobile services is to reconcile these previously conflicting values which are security and freedom.

1.4. Conclusion: "I am a nomad in at least five different ways"[36]

1.4.1. *A new challenge: reconciling the incompatible*

.Quitting the comforting environment of traditional fixed communication technologies – but it is already the case – in order to enter a world of new generation mobile services means in a certain way to be put in danger. This danger concerns the group of actors of the mobility, be them service providers or final users:

– loss of a controlled environment of an intranet or of a single network;

– mixing inside the same equipment – our cell phone – of historically separated usages between the handset and the computer;

– diving into a world where more universal sets coexist;

– higher fragility of users related to more and more requests;

– more need for protection within a more hostile environment.

1.4.2. *A combination of new technologies and new economic models*

Defining these dangers raises technological issues which started being designed together with the maturation of fixed Internet usage and with the multitude of RFC[37] that created the rich heritage of the IETF. They are replaced today by the 3GPP which, through the IMS, tries to extend them to vocal and mobile preoccupations at the same time. Such an evolution is to be found also between W3C[38] and the

36 Excerpt from *L'homme nomade*, by Jacques Attali.

37 The RFC (*request for comments*) are a set of documents referring to the Internet community that describe, specify, help the implementation, standardize and debate most of the norms, standards, technologies and protocols related to Internet and to networks in general.

38 W3C – *world wide web consortium*: http://www.w3.org.

OMA[39]. Nevertheless, there are many unknown issues and a playing field that stimulates many desires.

Thus, the migration of telecommunication networks – originally from the IN[40] model – toward the more and more intelligent and open terminals on the one hand and, on the other hand, the servers spread inside the global Internet network are a source for changing the way the cards are dealt. While some people thought that this trend was limited to the PC world, it is clear today that this evolution is equally unavoidable for the mobile services. The result is a redistribution of value between the old and new actors of this chain. It is also a source for new opportunities that are the results of the new challenges previously presented.

As for the final user, it is him anyway who makes decisions in a market economy. Hence, it is up to us, technology and services providers, to be able to meet his deepest aspirations and by respecting a reality principle that might have been forgotten during the years of the Internet bubble. The winners of adaptive services and usages will doubtlessly be those who will integrate the best Internet and mobile telephony practices in a perennial economic model.

---

39 OMA – *open mobile alliance*: http://www.openmobilealliance.org.

40 IN – *intelligent network*. An intelligent network is a telecommunications network where the intelligence is centered and separated from the commutation function.

Chapter 2

Object Modeling and Software-defined Radio

2.1. Introduction

In this first section, we try to outline a range of computer technologies that can contribute to the implementation of mobile radios. The reader will afterwards have an idea of the relationship between object modeling and software-defined radios. Our purpose is not to describe either object modeling or software radio in exhaustive depth, and we kindly ask our readers to forgive us for any shortcut or simplification we took the liberty to present here.

Software-defined radio covers many concepts. The main concept of this approach is the progressive replacement of the specialized hardware elements due to digital processing in the telecommunications equipment by software computations, done on generic processors. This approach has many advantages:

– the radio set quality can evolve by improving the algorithms implemented in the software, while a hardware version would require a change in hardware (or, more simply, a change in the whole equipment);

– ideally, the same material can be used to sustain different types of radio sets only by changing the software;

– in other words, the correct definition of an interface between hardware and software could develop these two components of a radio set independently. This

---

Chapter written by Antoine DELAUTRE and Yann DENEF.

boundary can be compared, on the one hand, with the one that exists between the applications and, on the other hand, with the platforms and operating systems.

Before going too far into object modeling, we consider it worth presenting a simplified version of the software development evolution in the last 30 years.

2.1.1. *History of the software industry*

The software industry is quite new. It is a self-sustained industry, which means that a significant effort with regard to software development aims to develop the software meant to simplify this industry:

– operating systems;

– new programming languages and related compilers;

– integrated development workshops, etc.

Processors are more and more powerful (Moore's law), software is more and more sophisticated, without the user of an application having the slightest idea regarding the improvements of that particular application. All these developments are guided by the same purpose: controlling complexity. More and more complex applications must be developed and controlled faster and more reliably.

In the last 30 years, this industry has been marked by several important phases:

– the standardization of the advanced programming language (COBOL, Fortran, LISP, C, etc.) which made the development of structured applications possible (1970);

– the introduction of graphic interface (1985) which provided the opportunity for non-specialized people to use computers;

– creation of new languages, object-oriented and modeling tools associated with these languages (1990). Introduction of "development frameworks" aiming to simplify the software development process;

– definition of system development workshop which integrates several modeling languages and introduction of software components (2000).

In relation to the subject we are interested in, the last two points are more important. Starting from the 1990s, the development cycle has been guided by the software controlled objects. Processings are always present but they are attached to their target objects. These are not independent functions anymore but methods associated with these objects. In addition, particular mechanisms integrated into the programming language (C++/Java) make it possible to control and even to forbid access to a method or to an internal characteristic of the object by another object. Hence, it is possible to improve the software by modifying some objects which form it in a more controlled way.

At the very beginning, object analysis and then the software development which is the consequence of this analysis, was presented to be the best solution for software development and maintenance, whose complexity continued to increase. The results were not in conformity with the expectations. The object approach could not solve these problems by itself. It was always possible, by using this new paradigm, to design software of bad quality, which could not evolve easily. In fact, nothing did guarantee, from this point of view, independence between the groups of objects used in order to offer a certain functionality. An object A could use objects B and C depending themselves on objects X, Y and Z and so on and so forth. The object method offered more rigor in the software breakdown but it did not impose it.

An additional abstraction level was introduced in order to solve this problem: the component level. Firstly, the component integration into software is not carried out as before, at the level of the source code, but at the level of the object code. Therefore, the component is an independent entity – application or library – made outside the functional context[1] in which it is going to be inserted. Its independence is thus guaranteed. The creation of data objects defined by their interfaces and the transformation applied on these data led to the elaboration of the component model. Since then, the different functions can be put forward by independent components accessible through their interfaces.

A software component is defined by its interfaces. An application can use only these interfaces in order to operate the component services. By using this advanced technique, it is possible to define a system by relying on basic components (which exist or are being developed) without entering into details concerning the component.

1 The functional context represents the occupation application. The technical context is the component's reception entity.

Using this model based on components, it will be possible to generate the software automatically from the model without writing even one line of code in a given programming language. The model is for the programming language which was, 30 years ago, for the machine language.

2.1.2. *Object modeling*

Object modeling makes it possible to represent a system in the form of collaborating objects. This system is perceived from different points of view:

– a static point of view, which simply indicates the relations between the objects that form the system. These relations can be of dependency, composition, aggregation or heritage;

– a dynamic point of view, which defines the inner processing effects realized by the objects or the processing effects distributed between the system's objects;

– a physical point of view, indicating the deployment of the objects on the equipment constituting the system.

Every point of view is realized by different diagrams. Regarding the static point of view, the approach of object modeling is a slightly Cartesian approach of the domain of interest in order to organize it in an arborescent form resembling a family tree which indicates the family and hereditary relations enabling the parents to recognize on their child's face different physical or emotional features of the family members. In fact, the search for the basic objects of a system and its organization is not unique but, on the contrary, it is subjective according to the purpose pursued.

If we take, for example, a piece of furniture called a "chair", it consists, generally speaking, of four legs sustaining a seat accompanied by a back support. An armchair can be described as an improved chair by the addition of armrests, while a stool could be explained as being a chair without the back support.

An object is defined through its own attributes and methods, i.e. the characteristics and operations which can be applied to this object. When the object hierarchy is created, it indicates the heritage in terms of attributes and methods.

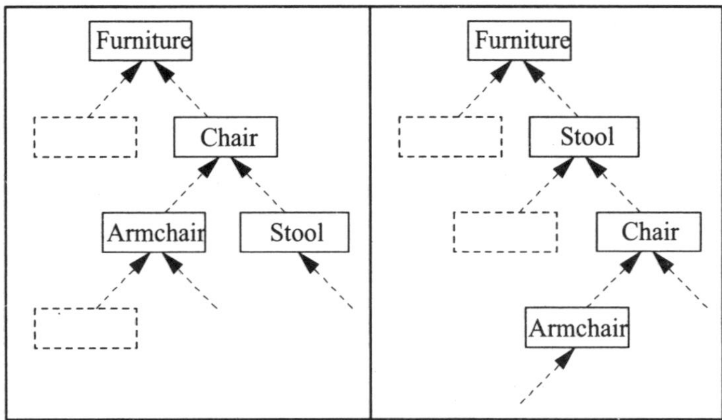

Figure 2.1. *Example of a hierarchy of objects based on two points of view*

Let us suppose that the attributes of a furniture object are: weight, length, width, height. By heritage, a chair will be defined through the attributes of the piece of furniture to which we can add, for instance, the weight it can sustain. In terms of method, we could define the fold and unfold methods which act on the object (camping) "chair"

However, a chair is composed of different elements: legs, seat, back support, glue, nails, etc. that we can call basic elements or meta-elements. A chair would be defined by selecting the basic elements and their composition. The basic objects depend on the field of activity: cabinet work or interior architecture. The choice of basic elements and their definition is an integral part of the modeling approach. The model of a system is not unique and it depends on the point of view considered.

2.1.3. *Modeling and data flow*

The modeling approach is an industrial approach of rationalization of the field in order to obtain the federative constituent elements around which other elements can be organized. It is often through the study of the data flow that we can identify which are the basic elements of a field. For the banking field, we generally consider the bank account as a central element. In fact, it is the structure entity which is used by the various contributors. The representation of the field requires a certain abstraction effort in order to identify the constituent elements which are used because often they are not physically real and they do not even exist in their initial form but only in an evolved and much more complex form: a bank account does not exist physically but it exists in its advanced form of the type: "current account", "security account", etc. and its existence is linked to the coexistence of other objects such as: holder, positive balance, branch, etc.

We can already easily imagine that this approach can be applied to the mobile radiocommunication field: it is an identified field which performs a complex processing of the data flow (voice) whose generic representation which we can make makes it possible to identify the main elements.

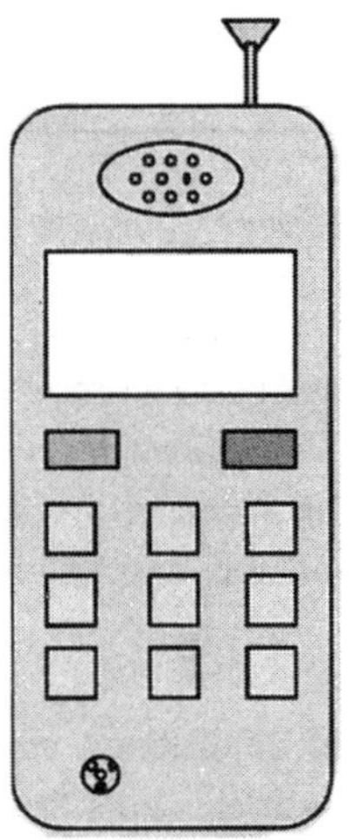

Figure 2.2. *Generic representation of the mobile radio*

The above representation, even though simplistic and caricatural, makes it possible to clearly identify the following physical constituent elements: microphone, keypad, display, speakers, antenna and casing. This representation also identifies the main input and output of the data flow: air vibrations and radio signals. The chain of processing mobile communication information calls for the common elements which are specialized according to the way in which the information must be processed: bandwidth, rate, accuracy, etc.

Information technology means processing information, i.e. transforming the data step by step. Computer programs are defined by the type of processing applied to the processed data. Methods, such as object analysis, have been created in order to formalize and to structure software analysis, based on data identification and on the transformations they go through.

2.1.4. *Constituent model*

The creation of data processing objects defined through their interfaces and the transformation applied on the data, led to the elaboration of a constituent model. This model tries to create an application in order to solve a problem by associating different data processing objects according to a well defined order and logic of the problem considered, while the different objects were individual created. This model has been created in order to simplify the monolithic programs and to make it possible to supply flexibility and progress by the independence of different constituents.

Figure 2.3. *Example of multiplication constituent*

This representation presents the main characteristics of the constituent model, namely: black box model, input-output connectors. The black box model does not provide any information on the selected implementation and hence it does not make any implicit or explicit hypothesis on the other constituent components of the application. The connectors identify the inputs-outputs and materialize the way of connecting the different components. A clear specification has to be associated with each component, explaining the interfaces' specifications and the processing of this component. We note that the coupling between different components is as loose as possible because it is reduced to the exchanged data. To make a constituent function, it is necessary to implement an operating system that is adapted and to offer management mechanisms for the life cycle of these components as well as mechanisms favoring their interconnections.

2.1.5. *Software bus*

Since computers have to be administered and controlled when they are interconnected in the network, a remote control mechanism is implemented. The initial principle is based on a client-server type mechanism. On each machine that needs to be administered, a server is run. In the world of the Internet, a server is a special program waiting for the network connection. The connections are initialized by the client. The client has to know the address of the server (IP address) and the destination port of the server. Each service offered is characterized by a special number[2]. Numerous services usually use this principle and they are available in Unix environments:

– reliable file transfer (ftp);

– offline file transfer (tftp);

– remote order processing (rsh);

– remote file copying (rcp);

– remote session processing (telnet or rlogin).

At the beginning of the 1980s, Sun introduced an additional abstraction level between client and server: remote procedure call (RPC). The remote services have always been implemented by servers, but the servers never responded to a certain port number. The services are always identified by a number but this number is not the port number. A separate server (the port guide) ensures the correspondence between the service number and the program to call. In addition, the integration between client and server is more direct because from a programming point of view, the client calls the server in the same way as he would make a local call. All the aspects of network communication (socket opening, connection, etc.) and data format are hidden to the programmer. This RPC mechanism is, among other things, used for remote disk sharing through NFS (*network file system*).

The RPC mechanism was extended and generalized by OMG (*Object Management Group*) in Corba's definition (*common object request broken architecture*): a service or a function is presented and made independent from the machine, the operating system or the programming language used. In addition, complementary services of localization, distribution, subscription, etc. are defined. The COM/DCOM™ environments and Java/RMI or JINI are the avatars, addressing a part restricted to an operating system or to a universal computer language.

When using Corba, the information exchanged between two components passes through Corba which is in charge of translating it into an intermediate language of fixed representation in order to overcome the particularities of processors, languages and operating systems. With this aim, the IDL (*interface definition language*) has been established and the correspondences with different languages defined (C, C++,

2 This number is a TCP port number (connection-oriented server) or a UDP port number (connectionless sever).

ADA, Java, etc.). Corba's purpose is to "make a program believe" that the programs to which it is in relation are collocated on the machine on which it is being run and written in the same language as illustrated in Figure 2.4.

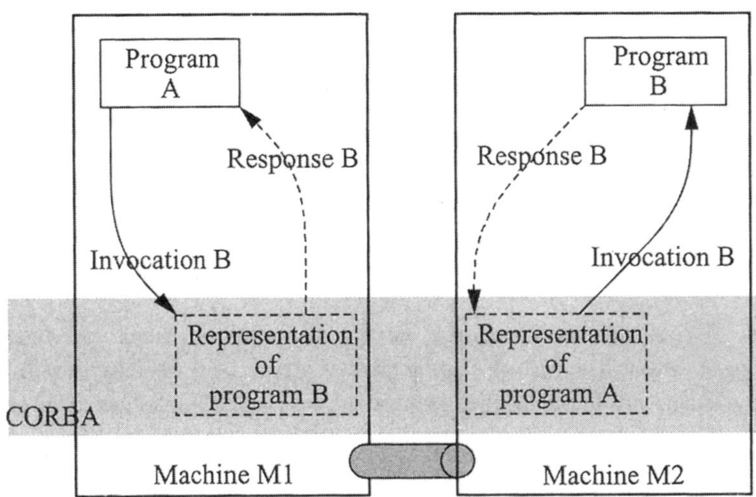

Figure 2.4. *Illustration of Corba's operating mode*

When program A is launched, a representation of program B or *stub* is created by Corba. When program A performs an invocation of program B, it is the representation of B that is invoked in machine M1. This representation translates the invocation into an intermediate language, then Corba determines the target machine and sends the necessary information to it; if necessary, a representation of program A or *skeleton* is created. If need be, the response is sent by the reverse path.

Complementary services can be implemented in order to locate the programs during their invocation, to favor data broadcast, to subscribe to data, etc.

The mechanisms linked to Corba can be static or dynamic. In the first case, the client knows the component interface and uses this interface (*stub*) during compilation whereas in the second case, the client discovers this interface during the execution by calling the services proposed by the *repository* interface.

In any case, the composition between the various components necessary to realize a computer program is completed when each component invokes another one. This "real-time" mechanism makes it possible to take into consideration a new functionality, such as a database or an algorithm or a sensor, as soon as the associated component is posted or registered in the system. If the component interfaces are not modified, the registration of a new component version makes it possible to improve the client application without having to do a recompilation or a new editing of links.

We can easily notice that Corba's mechanisms are very well adapted for the centralized computer systems and make it possible to control the load distribution or the redundancy. They also enable the easy creation of industrial process management through the possibility of interconnections of different and various sensors of different providers and the centralization of information produced in a database.

The services offered by Corba intrinsically cost in terms of delay and computation power. The software performances are variable because they find in a generally sharp way the colocalizations or the languages implemented in order to diminish the translations at the maximum and to simplify the inter-machine transmissions. Similarly, certain implementations reduce repeated copying or make it possible to take into account the patented bus with the purpose of improving the processing to be done.

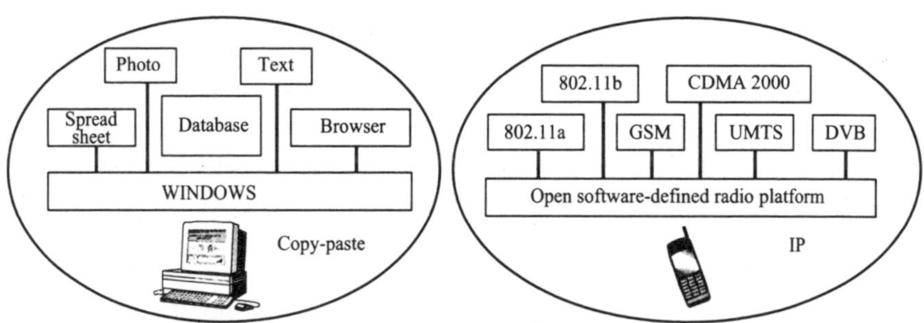

Figure 2.5. *Analogy of the suggested radio software approach*

Creating components with no interdependence makes it possible to create on-demand applications, similar to a construction kit. To do this, it is necessary to implement a common home interface consisting of basic services provided by the platform which will install, deploy and interconnect the components, in order to realize the desired application.

2.1.6. *Product line*

A product line approach has many advantages for a software editor as it makes it possible to decouple the application from the platform. A major industrial consequence is the decorrelation of software and hardware product plans in temporal and functional terms.

In the same way, the different components which form a service can evolve independently. Each element can follow its own life cycle without having an impact on the other elements. It is also possible to realize components common to many products or to supply the market. The manufacturer can concentrate on his activity and buy the technologies that are common or non-controlled on the outside. If an element that is exterior or from the past is not adapted to the platform, a "wrapper" is created in order to receive the element and to present it to the platform as being a component which respects the model requested by it.

The creation of the component-based architecture requires, for the field we are talking about, to have evolved to a sufficient phase of technical maturity so that the architecture is perennial to the minimum, making the creation investment of the structure and of the platform environment profitable. This is a difficult approach for a company both from the technical and financial points of view, for it imposes a *de facto* reorganization of the company: identification of key activities of the company, cancellation of certain sectors. The resulting online organization of products describes the inner structure of the company and the dependency links with the provider companies.

If we take, for example, our personal computers, they are organized around a communication bus initially created by IBM™ and named PC AT™. Irrespective of the improvements to this bus, the architecture is the same. Many providers took advantage and created products for certain functions (disk, graphic card, etc.) for the common benefit since computers improve in terms of performance and functions for an average price which has been the same in absolute value for the last 20 years (€1,500). The initiator of the approach, IBM™, is both an integrator of personal computers and a manufacturer of components (hard disk, for instance). New companies were created, the components improve in their own rhythm and a large number of users can upgrade their computer by parts.

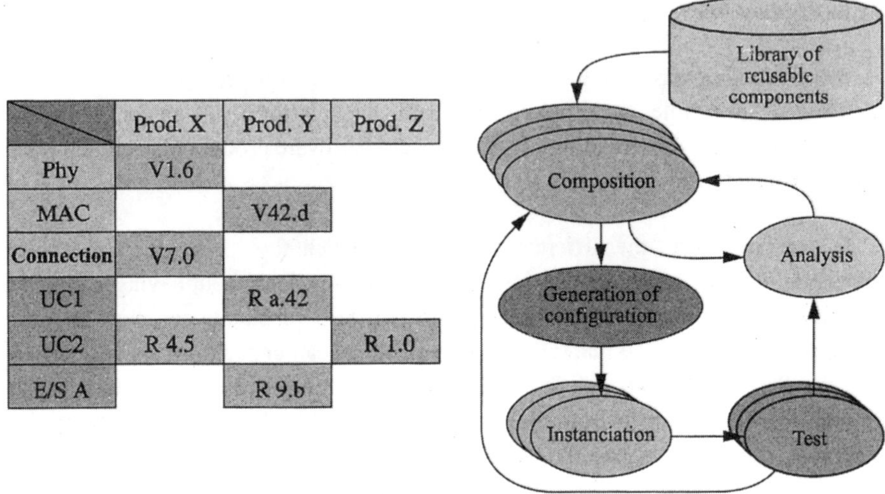

Figure 2.6. *Products line based on the component approach*

The research and development investment, still called non-recurrent cost, grows with the complexity of a monolithic product and therefore it can cause a financial stability problem to a company. To reduce this investment or, at least to minimize its progress, the developers try to "freeze" some functions or to "parse" them into loosely coupled elements. This attitude can appear to be very good with respect to the company's investments, but unfortunately the consequence is sometimes to freeze the structure of the product, which means its architecture, without knowing or deciding the line of the related product. In the absence of a defined and coherent product strategy, the technique will decide its own criteria.

The automobile is another field of activity which is organized around platforms that are sometimes common to several models or manufacturers. The motor is generally divided into several cylinders based on the same "engine block" and it can be common to many models. The manufacturers have also become the integrators who order components from contractors. The mutation of the automobile became possible through the evolution of the technologies used but it was also required by the need to reduce the time necessary to launch it on the market. Today, when a new model is created, a manufacturer digs into a "basis of components", which are produced at low prices and which are already reliable. The time needed to define the model is reduced and evolution possibilities are left in order to adjust the model to the different market needs and to their temporal evolution. Hence, the objectives of adaptability, flexibility of evolution and reliability are accomplished.

2.2. Applicability of the component-based approach to the field of software-defined radio

The component-based approach has many advantages but it is on a platform and imposes an evolution of the products, i.e. a consequent development effort. It is difficult to evaluate the effort necessary to create these products because it depends on the maturity of the company and on the relation between the components bought and those produced by the company itself. In the rest of this chapter, we would like to demonstrate that the component-based approach is very well adapted to the software-defined radio and to the technical and economic context, but we will also show which are the implementation difficulties and the technical additional costs.

2.2.1. *Software-defined radio*

First of all, we will concentrate our presentation on the software field in the broad sense of the word by giving a definition of software-defined radio:

> "*A software radio is a radio whose channel modulation waveforms are defined in software*" (Joe Mitola[3]).

This definition is considered standard because it is one of the first definitions of software-defined radio and it describes the need expressed by users or professionals who are interested in this subject. The objective researched is to be able to use the same hardware for different communications, for instance DECT and GSM; the terminal specialization is done by the software.

As we will show in the following chapters, the progress in digital conversion and in data signal processing made possible and even more efficient the processing of information by software. Memories improve both in size and speed, making it possible to upload the whole code into the processor's memory, FPGA can be re-programmed, the converters and different generators of pulses can be programmed on demand. Therefore, it becomes a realistic option to specialize an industrial product through software in order to realize the desired waveform with cost

3 "Software radios: wireless architectures for the 21st century", http://ourworld.compuserve.com/homepages/mitola.

performances, satisfactory consumption and weight. However, hardware designers currently create their own software implementation in order to obtain a competitive product; only the "high" or protocol layers are relatively independent of hardware.

The concept of software-defined radio must make it possible to successively implement several waveforms according to the user's need: an instant communication, an implementation process, material resources available or an electric radio spectrum status. Therefore, it is important to offer many update or evolution possibilities in order to improve the communications capability: either in advance by anticipation or during utilization as a certain need arises. In fact, if some ways of communication are well defined and stable (GSM, DCS), others are much more adjustable and variable (802.11 – Wi-Fi).

Furthermore, added value services are offered by operators along with their creation and success (SMS, chat, games) and lead to updates of the applications or of the communication protocols. The concept of software-defined radio is the opposite of a "disposable" hardware bought in order to be used in a given context and then renewed by another product 6 or 12 months later. It must be perennial and it must guarantee its adaptability to the named evolutions through downloading and adapting the software.

For the telecommunications operator, real or virtual, and for the user of a "fleet of terminals", the software-defined radio must be able to provide a homogenous administration of the fleet of terminals: both for the updates and for the user's available and distributed versions, as well as for the distribution of advanced or targeted functionalities for a job or an activity.

Even though desirable from the user's or operator's point of view, software-defined radio is submitted to technical constraints. The favored way of updating or of configuring terminals is, of course, the radio channel. However, it generally remains limited in terms of throughput, some ko/s; hence, it is preferable not to transmit large amount of data. The use of the radio channel is the billing source of operators; its use for non-billing operations is limited to the minimum or transferred to moments of under-loading of the network. The memory capacity of the terminals, even if it grows fast, remains limited in terms of applications, so it is advisable to load only the necessary data. Finally, there are many terminal and access point manufacturers who offer many products, regularly renewed. The number of possible associations between terminals and access stations is very high thus leading for sure to errors or operating anomalies; it is anticipated to solve these anomalies by updating the faulty hardware.

This is why it is not considered to create a software image through possible configuration, but to realize the software image only through the combination of different modules with the purpose of passing on only the faulty module.

2.2.2. *Evolution of the industrial tissue*

The evolution of technology favors the implementation of much more competitive communication solutions in terms of emitted power distribution and encoding in order to obtain higher throughput, but this can be realized only at the expense of a better precision, of a more complex processing and of an increased processing capability. The software part has thus increased from 5 to 10, from 100,000 lines of code to 1,000,000 lines of code. Therefore, investment becomes much more significant for manufacturers, who have to invent, develop and industrialize the hardware, all at the same time.

However, the success of broadband communications, which would justify the use of these new networks, cannot reside in voice transmission and has to be stimulated by new services. To balance this challenge, manufacturers will improve the architectures in order to incorporate hardware or applicative components: antenna, battery, software (Flexi Matrix, Hughes Software Systems, AudioCodes, Motia, Symbian, Sacet, etc.).

The increase of the complexity of terminals and their management generated the creation of companies that developed distribution or network asset management solutions for software updates with regard to a fleet of terminals (Swapcom, NamoDigit Corporation). These societies offer suitable solutions that meet the need for a homogenous management of terminal fleets for the operators through patented federative solutions in conformity with standards.

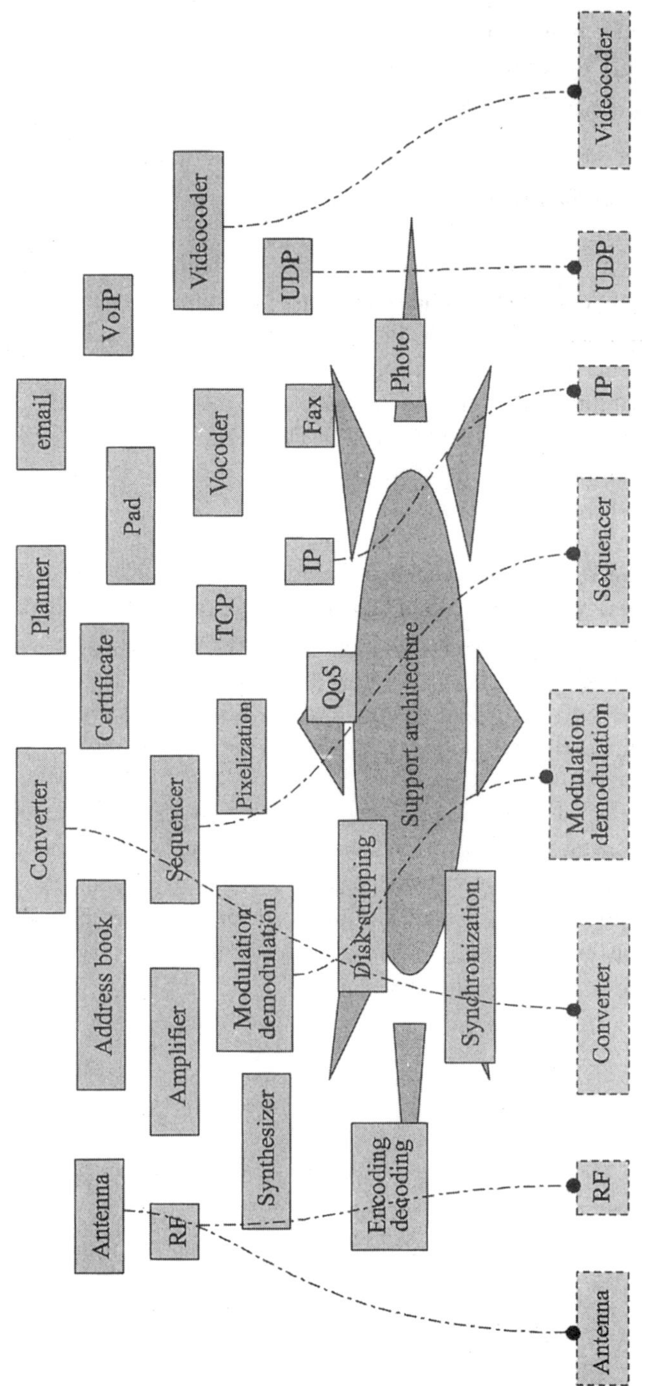

Figure 2.7. *Structure of modules necessary to create a radio processing chain*

2.2.3. *Need for stable interfaces*

Nowadays, the technical field has to face the development of new protocols (UMTS, 802.1xy) and needs many standardized protocols (802.11, Bluetooth, GSM, etc.), whereas the manufacturers find themselves in front of the increasing complexity that generates new actors offering services both to operators and to manufacturers or integrators. The conditions of a technological break leading to a new organization of the field seem to be now brought together.

In order to be able to answer the development of all these protocols, the actors of this field have to agree on a common structure capable of constituting the basic architecture on which all products can be assembled. The only existent computing structure corresponding to the aforesaid criterion is the component-based approach. It is impossible to impose to each manufacturer to use exactly the same model with the same interfaces, but it is important to identify the same principles and the same main functions.

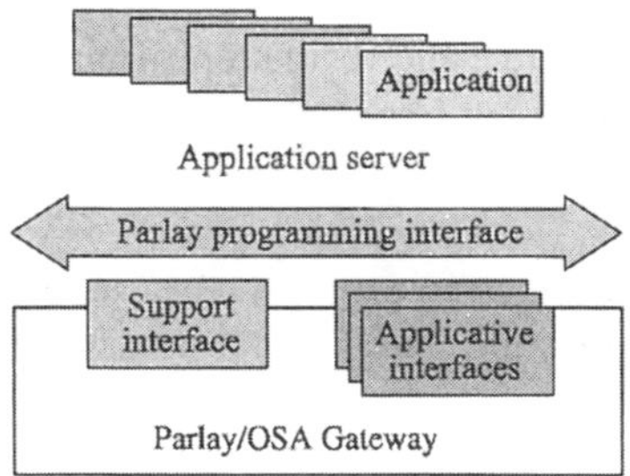

Figure 2.8. *PARALAY/OSA model architecture*

It is not possible that each manufacturer offers its own complete model and management tools to the operators for it will be unmanageable and in conclusion useless. A minimum core must be defined, making it possible to manage the terminal configuration according to the context of use. On the other hand, it is necessary to propose standardized interfaces and common environments in order to enable third party companies to propose hardware and software components that can be used in different terminals in order to reduce the development costs. The initiatives, such as Classmark MeXe or PARLAY/OSA, define the zones of execution opened by the manufacturers for executables provided by third party companies.

2.3. The constraints of the component-based approach

The component-based approach imposes design constraints on the software because it has to be broken down into modules that should be independent or loosely interconnected by data streams. This means that the order of processings to be done has to be sent through data.

2.3.1. *Execution time constraints*

The planning of a program is based on one of the following three factors: time, event or data. The physical layer of a communication protocol is generally organized on a temporal rhythm which can vary according to events: communication slots and temporal dates, the content being encoded based on a diagram according to the transmission quality. The superior layers, connection and transmission, are usually sequenced according to the received data or to the data to be sent out. These high layers are easily reorganized or modified in order to become components.

As an example, we can quote the regulation of power emission for the UMTS protocol which imposes a reaction less than a second between the reception and the emission of the order.

To show how difficult it is to transform a physical layer into components, let us consider the disk striping function: this function can be realized only if a consistent group of data was sent, no longer according to a moment or a discontinuation, both being dedicated to an execution platform. In the same way, it is impossible to make predictions on the order in which the processing will be executed since neither the processor nor the scheduler of the operating system are specified (the organizing logic is not standardized). Since the load of the processor is not stable because it is linked to user services or to the updated protocol, the execution order of the processes or the data issued from the processors cannot be determined or considered stable. These different constraints imposed on the system and component design lead to reiterate the software design in order to make them reliable and to "decouple" them from their environment. A short term possibility is to "freeze" the existent software into big components, corresponding to one or several protocol layers.

2.3.2. *Software – hardware coupling constraints*

We know that in the context of software-defined radio, the software does not "know" the platform on which it is run: processor speed, memory capacity, dedicated coprocessors or components. The only knowledge of the platform on which rests the component is a list of services provided: real-time operating system, file management, etc.

These services are defined for all platforms of software-defined radio and therefore, all software must be adapted in order to use only the common services instead of "patented" services, such as a memory address an input/output port.

Terminals are produced serially favoring the creation of dedicated chips in order to minimize the manufacture costs and the power consumption and to maximize the period of operation between battery chargings. Many dedicated functions, simple but calculation expensive, are integrated in the form of hardware accelerators in an ASIC. In the case of software-defined radio, these accelerators can be used only if they are distributed in all the software-defined radio platforms.

The first mobiles and the first basic stations were realized by pushing the existent technology to its limit: the software was custom-made for every piece of hardware, combining the designs of hardware and software. This made it possible to realize efficient products, but their adaptability and their evolution are still limited.

The design constraints for platforms and software are too strong in order to reach the flexibility required by the software-defined radio, but they are already implemented in the general or management computer technology. The appropriate techniques exist and have to be adapted to the target environment. The resulting generality is not synonymous with uniformity; the patented technologies can always develop both in hardware as in software. If the patented technology corresponds to a coupling between a particularity of the hardware and a particularity of the software, it is during the assembling phase that the particularity of hardware and software configurations will have been processed, favoring specific processing.

2.3.3. *Reminder on the evolution of software technologies*

Java is an example of the success known with the approach of a common standard platform on which software is run independently of the real hardware platform. It is obvious that the performances are not the same according to the underlying hardware but real-time applications such as video display have been written in Java (DVD player).

Just like in the case of Java, the component-based approach leads to additional costs at the level of the processor and memory because the platform which integrates the operating system and the basic services requires a part of the computation power and of the memory. This additional cost which makes it possible to gain in terms of flexibility and reuse is not unacceptable; the embedded computer science has already known this kind of evolution in the past. Let us recall the main transitions in this field: the passage from microcode to the assembler and then to "evolved" language like C, the introduction of operating systems, the introduction of object languages such as C++, the introduction of Java and use of operating systems such as Unix (real-time Linux). Every evolution needed an improvement of the performances of the platform implemented, but it also made it possible to improve the productivity of software development to the overall benefit of the platform. These evolutions also made possible the creation and implementation of high level complexity software like the protocol stacks GSM, 802.11 and UMTS. Since the component-based technology has already been adopted in many information technology fields (management, air traffic, finance, etc.), there is no doubt that this technology will successfully be adapted to the requirements of software-defined radio. We note for instance, the appearance of dedicated software bus adapted to the signal processing environment with Vetel eORB for the DSB target.

2.3.4. *Regulatory constraints*

If, as we have just presented, technology offers solutions for the creation of software-defined radio, we have to keep in mind that the terminals and the radio access points are the elements submitted to regulations. This regulation varies from one country to another. First of all, the transmission power and the frequencies used are governed by local rules; however, it is the manufacturer or the operator who is responsible for the rule application. The user cannot be held responsible for the malfunction of his equipment because he is very rarely aware of the application rules and he does not have the technical means to verify or modify the functioning of his equipment. This responsibility implies the introduction of control, safety and security mechanisms into the equipment, in case of wrong or dishonest handling or simply in case of country change. Currently, all pieces of equipment are certified for use in the country where they are sold and an identification number makes it possible to know its hardware and software structure.

Software-defined radio cannot be certified in advance because the operating mode and the waveforms have not yet been created or at least implemented on the equipment. The hardware must provide self-protection mechanisms in order to meet the safety rules, irrespective of the waveform which could be configured and used. Connected to these mechanisms, it has to be possible for the regulatory authorities of the spectrum frequency or for the operators to "reject" an equipment whose operating mode is dangerous for the user or disturbing for the good operation of the radiocommunication network: in such a case, the equipment has to offer, however, a minimum communication service, which can be called a basic service, enabling emergency calls and voice transmission telecommunication services according to the user's contract. Since communication equipment has become a link in our individual security chain, it has to offer emergency services which will surely extend in the case of software-defined radio: apart from "999 and 991", the transmission of a distress beacon on the VHF or HF channels toward the police department or firemen. The flexibility of the software-defined radio will make it possible to reduce the uncovered radio zones for hikers, mountaineers and professionals in rural areas and hence it will have to improve its reliability and user-friendly attribute.

2.3.5. *Deployment constraints*

We can consider different ways to (re)configure software-defined radio. The first one of them consists of a CD-Rom and a line delivered along with the equipment which makes it possible to parameterize and configure from a standard computer. A second solution, derived from the first one, can be done via an Internet connection toward the manufacturer or operator. This will make it possible to adapt the operating mode of the hardware according to the anticipated use. The user's preferences, his profile and the terms of his contract with his operator(s) can be done in such a way as to generate the management rules for software-defined radio. However, the easiest solution will be a configuration at the moment of purchase, made by an expert, or remotely, during a call to a support center that will perform the remote update.

In this first configuration step, it seems to be necessary for the operator to be aware of the terminal capacities in order to know its possibilities with respect to the possibilities of the contract concluded. Let us take the example of a user traveling for fun all around the globe and who has a reconfigurable terminal. Starting from France, he uploads by default in his terminal the modes used in France (GSM, UMTS). However, when he reaches the USA, he can also upload additional modes (for instance, CDMA 2000) if his terminal accepts this operating mode. When this user wants to organize a trip, he will ask for an adaptation of his terminal configuration so that the terminal should be operational when he arrives at destination or during his trip. During this preparation, the user can call his usual operator or can choose another operator during the duration of his trip. Moreover, a new configuration could be established, taking into account the travel steps in order to guarantee continuous connection.

If the first configuration can be done by an expert during the purchase of the terminal, the following configurations will be at the user's initiative or the network operator. The final control of a reconfiguration should be probably done by the operator in order to guarantee the reliability of the reconfiguration. The operator, together with the terminal manufacturer, will activate a tested or at least evaluated configuration, will keep track of the operations performed and will remotely test the new terminal configuration. This notion of responsibility in the reconfiguration management of mobile terminals reconfiguration is one of the main dampers toward a large deployment of reconfigurable mobile radio networks.

The case of basic stations is much easier because they are considered fixed and teleoperated by experts. Their shifting toward the technology of software-defined radio will be quite easy. In this case, the difficulty will reside in the choice of protocols and frequencies that enable a satisfactory balance between the number of connected subscribers, the throughput charge per subscriber and the size of the bandwidth.

2.4. An outline of the works pertaining to the component-based approach for software-defined radio

2.4.1. *SPEAKeasy and JTRS*

The Department of Defense of the USA initiated works on software-defined radio in the 1980s. At the end of the 1970s, in the air force research lab (AFRL) the prototype "*integrated communications navigation, identification and avionics system (ICNIA)*" was developed. ICNIA can be considered as the predecessor of software-defined radio because it was based on the use of digital signal processors (DSP) and general processors (GPP – general purpose processor). The success of this prototype gave way to the development of a more ambitious program, called SPEAKeasy; the objective was to develop a multichannel (4) and multiband (HF, VHF, UHF) prototype. The SPEAKeasy prototype was presented during the interoperability exercise of June 1995, the JWID-95 (Joint Warrior Interoperability Demonstration). The SPEAKeasy program can be considered as the first research program which demonstrated the possibility of creating a multichannel multiband radio, reconfigurable by software.

2.4.2. *The weight of the USA*

The very encouraging results of the SPEAKeasy prototype were used in a much more ambitious industrial program, the JTRS (joint tactical radio systems, http://jtrs.army.mil). The JTRS program has a $6.8 billion budget, all technological domains put together. The basis for the decision of launching the JTRS program lies in the external operations of the US army: "Grenada Operation" and "Desert Storm", which demonstrated the limits of communication systems deployed in terms of bandwidth and usage flexibility leading to a saturation of satellite communication systems and the use of civilian communication systems by using prepaid cards. It also has to be taken into consideration the position of the Department of Defense of the USA which is at the same time the user, the access provider and prescriber. At the end of the 1990s the military radio fleet of the USA had 750,000 pieces of equipment for 200 different models and 40 waveforms: hence a very complex and expensive logistic system is necessary as well as dedicated systems for the communications between forces.

This quite difficult situation of the Department of Defense of the USA at the end of the 1990s takes place in a period when the predominance of the voice disappears, giving way to data and Internet. It is necessary to develop radiocommunication systems in order to take into consideration the data transfer and to increase throughput. All these technical considerations and certainly others led to the Department of Defense of the USA publishing on August 21, 1997 the "mission need statement for the joint tactical radio". The objective was to create a reduced number of hardware platforms able to receive the waveform protocols compatible with all platforms. The platforms are made of hardware elements whose specifications are standard and vary depending on the usage environment (airport, sea, etc.).

2.4.3. *The impact of JTRS on industrial sector technologies*

It seems necessary to mention the JTRS program as it is a characteristic example of policy strategy change of a client in favor of a technological breakdown. On the other hand, this program finances different technologies related to software-defined radio, like MEMS (micro-electro-mechanical systems) or broadband amplifiers or adaptive components (adaptive chip) and thus it directly influences the state of the art. Another characteristic of the JTRS approach is the wish to standardize the components in order to decrease the acquisition costs and thus the overall ownership costs. To be able to standardize, JTRS started a campaign of sensitization and promotion of software-defined radio. Hence, a great number of communities and universities got together in this campaign through contracts or forums (SDR Forum, OMG, WWRF). The financial and industrial power of the JTRS program is such that the promoted ideas will certainly orient the development of the next generation of military and civilian products because the products and standards issued from the JTRS program are launched on the market by its own will, in order to favor the adoption and to minimize ownership costs.

It shall be noted that the military context of the USA is present in many countries because the needs in communication and data transmission are no longer met by the current radio equipment. Recent conflicts or peace maintenance operations done by the coalition forces of different countries lead to the need of intercommunication. The same context exists in a simplified variant at the local level, such as intervention and security forces like firemen, registrars, policemen, ambulances, doctors, etc. In this last case, the civil communication network, which serves as a relay, becomes vulnerable to overload, to attacks and bad weather, and it does not cover the area. In France, a winter storm or a significant flood dramatically disturbs the forms of communications making the cooperation of different state sectors hard.

The importance of the JTRS program may look weak in comparison to the various research programs initiated in the radio domain, but we must take into consideration that this program oriented by a single client into a preset direction. As a comparison, the research initiative of the European Commission addresses several domains and it is of a cooperative and federal type, no longer oriented toward a single client.

2.4.4. *Communication software architecture*

In the previous sections we gave an important place to the JTRS program because it is at the center of the approach and proposes a standardization approach in the OMG – *Object Management Group* (www.omg.org, sbc.omg.org). The initial frame of the JTSR software-defined radio is described in the SCA document – *software communications architecture specification*.

Figure 2.9 sums up the ideas of the JTRS software architecture which is composed of components connected by a Corba software bus. The military domain imposes strong security measures leading to a separation between the so-called red or sensitive data (clear data) and black or protected data (coded data). Every protocol layer is identified by a dedicated interface MAC, LLC (*logical link control*), Network defined in an interface definition document. The "CF – core framework" or central structure gives all the necessary services to install, initiate, configure, start or stop a protocol.

The platform is itself defined as a conceptual model of material resources, an object-tree type structure which is able to be associated with a component-organized application that shares the material resources.

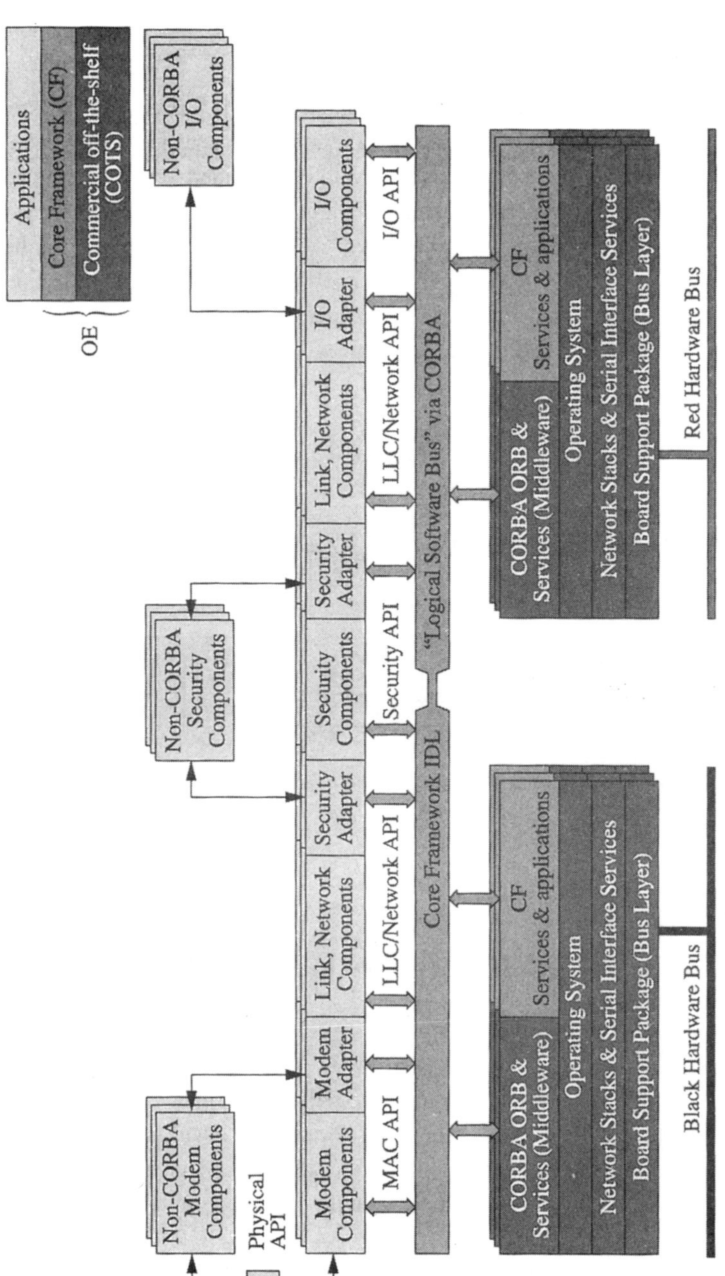

Figure 2.9. *Software architectural structure – from SCA rev 2.2*

2.4.5. *Hardware architecture*

To each hardware or software resource is associated an XML (*extended mark-up language*) label that indicates its requirements and characteristics. This label indicates the interfaces provided, i.e. those that are to be provided at input and also the memory requirements or the target processor. Figure 2.10 gives a general view of the XML model of the hardware architecture. A hardware platform will be described in the form of an XML folder pointing itself to every XML resource file. During the installation of an application, the *core framework* will verify step by step the suitability of the application and of the platform.

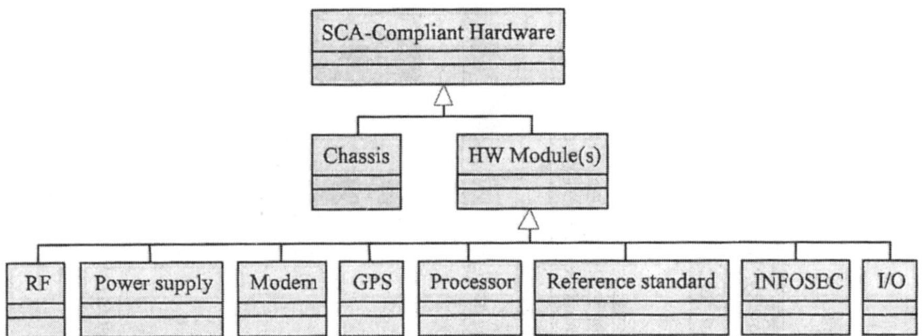

Figure 2.10. *Hardware architectural structure – from SCA rev 2.2*

2.4.6. *Standardizing activities*

The different concepts and objectives of the SCA were made popular and discussed in the scientific community as a favored subject through the SDR Forum (Software Defined Radio Forum, www.sdrforum.org). This forum is an independent organization created in 1996 with the aim to promote software-defined radio in the civilian, commercial and military areas, bringing together more than 100 companies from all over the world. The objective of the SDR Forum is to develop the rules and standards of software-defined radio, to facilitate the adoption of open architecture for radiocommunication systems and to cooperate with the regulatory agencies.

In its endeavor to promote the software-defined radio and in order to contribute to the adoption of the SCA concepts adopted by the SDR Forum, a development initiative in free software of the SCA was supported and perfected.

The SDR Forum was not a regulatory agency and therefore the SCA actors turned toward the OMG in order to regulate their work.

The OMG is defined as the greatest software consortium in the world; created in 1989, its mission is to help software users to solve their integration problems by providing open, neutral, mobile and interoperable specifications. OMG is a regulatory group consisting of 800 military or commercial companies, which managed to develop technologies like Corba (*common object request broker architecture*) and UML (*unified modeling language*), to quote only the best known. Its defined and structured organization enabled the creation of a great number of workgroups on different subjects such as space, health or sonars, which elaborate APIs (*application programming interface*) and then regulate them. The big telecommunication companies have OMG representatives who actively contribute to the regulatory process.

Among the OMG activities, some of them have a direct impact on the software-defined radio, i.e.:

– deployment and configuration;

– easy Corba component model;

– software based communications.

The "deployment and configuration" group works at the description of the necessary execution infrastructure distributed in order to make autonomous the *deployment and configuration of component based distributed application specification*. The "easy Corba component model" group has set as its objective to reduce memory constraints and calculation resources as well as the memory print program implicit to the implementations of the Corba component model, with the aim to address real-time embedded systems such as radiocommunications or automobile. To do this, the group reduced the functionalities of the component model (*lightweight CCM*). Finally, a dedicated workgroup, the SBC (*software-based communication*), takes into consideration only the problem of software-defined radio in the broad sense of the word. This group was initiated by members of the SDR Forum who wanted SCA work to be regulated in order to be able to elaborate products on this basis and to develop the work on this subject.

2.4.7. *UML profile for software-defined radio*

The SBC group created a complete model of software-defined radio, described by the document "PIM and PSM for SWRADIO". This document included a UML profile for software radio. A UML profile is a UML technique for a particular work domain, whose purpose is to define the basic objects of the work domain which can be manipulated by an architect in order to build a work application. SWARDIO profile describes the basic elements of a software-defined radio, as well as the waveform components and the environment consisting of *radio devices*, like an amplifier or a converter, components of radio management and operating system.

The SWRADIO profile is organized in UML package: *applications and devices* defining the control interfaces of radio specific hardware components (SWRAPI – *software radio API*), *infrastructure* defining the deployment and architecture services of an application such as a waveform, *communication equipment* defining the logic model of the support hardware equipment and the description methods of associations usable by *infrastructure*.

2.4.7.1. *Resources metamodel for software-defined radio*

The resource is the central element of the application and must have interfaces that make it possible to identify it, control its life cycle, to configure it, to interconnect it to other resources through "ports".

We notice that resources can be tested and access to hardware is limited to SWRAPI and to the operating system. Waveforms or the application use the communication ports as links between resources or components; it is the resulting structure which provides sense to the waveform. It is possible to determine all the dependencies of a resource with respect to others as well as the use of the radio platform (SWRADIO). The flexibility, the independence and the generality of the solution respond exactly to the aspect of software radio presented in this chapter.

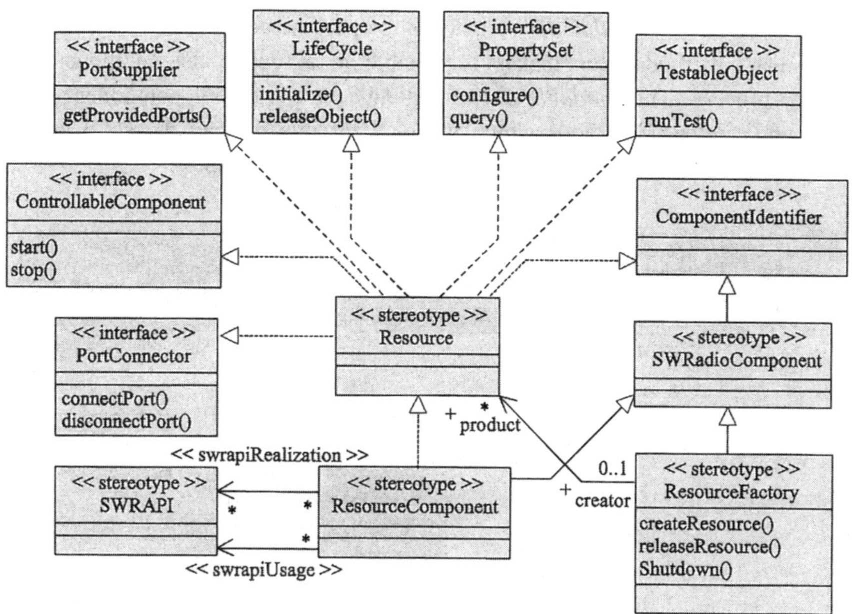

Figure 2.11. *Resource metamodel for software radio – excerpt from PIM and PSM for SWRADIO*

2.4.7.2. *Model of peripheral component*

The peripheral component or *device component* is the main concept used in order to create an abstraction of the implementation support hardware. The peripheral component is a form of *resource component* and *manage service component* with additional properties. It is a component whose usage can be known and it can reach its administrative and operational state. In order to be sure of certain capabilities of the peripherals it is possible to obtain their characteristics (memory, performance, etc.). Since the technical characteristics of peripherals are variable, they are described in the component descriptor. When the peripheral component consists of one or several other peripherals, the composition is indicated via *composite device*. The current name of the peripheral component is obtained through *labels* which also can indicate the location (for example, audio 1, serial 2). The object methods provide request and capability allocation mechanisms of the peripheral component or mechanisms that release or remove the capability. It is also possible to release the peripheral equipment (*release object*) or to release all underlying resources (*composite device*). This logic disassembly of the resource opens the possibility of composing a new peripheral component for a different need or to perform a physical disassembly. The *executable device* extends the peripheral component by "execute" and "finish" methods; this makes it possible to associate the component to a software process. The *loadable device* extends the peripheral component by the "load" and "unload" methods which make it possible to load the software on a peripheral component in accordance with the associated descriptor.

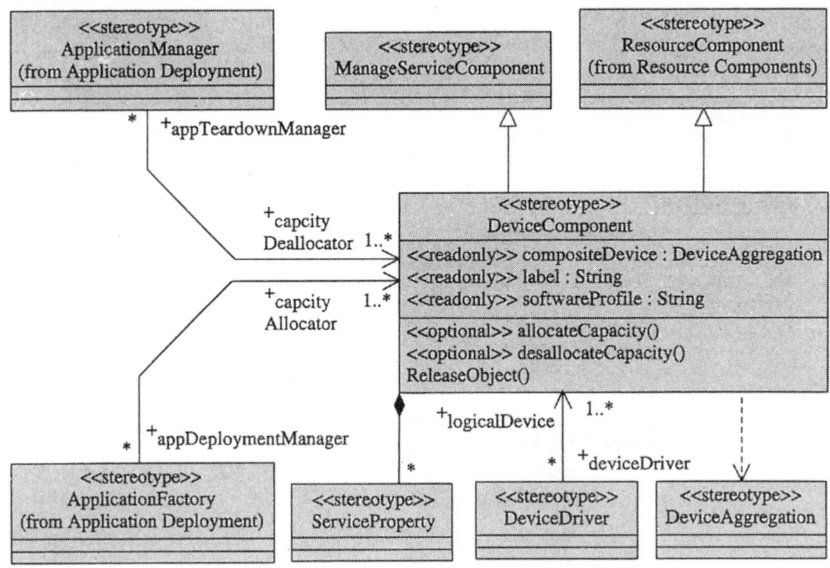

Figure 2.12. *Model of device component – excerpt from PIM and PSM for SWRADIO*

2.4.7.3. *Communication channel*

The UML profile also defines the basic elements that make it possible to describe a communication channel peripheral as well as the relations between peripherals. These definitions do not impose an organization of the software radio or a restriction of the operation domain which are left at the choice of the integrator or manufacturer. Their aim is to define a language that can describe the hardware platform of the software radio. The description of the platform can be registered into an XML format folder for an automatic analysis.

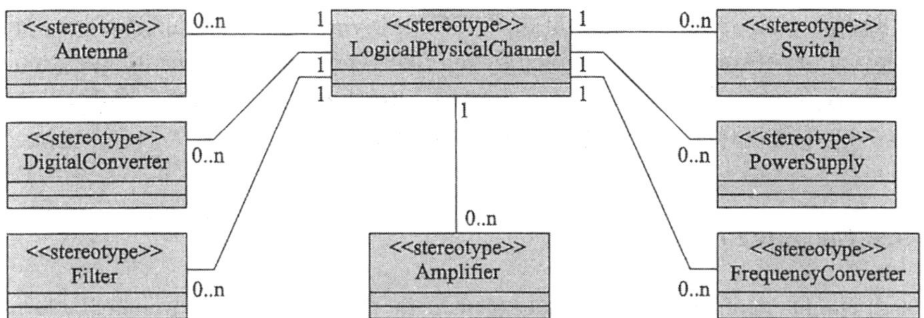

Figure 2.13. *Definition of logical physical channel – excerpt from PIM and PSM for SWRADIO*

Figure 2.13 shows the manner in which a logical physical channel is defined. This channel will be associated with two logical channels: communication, secured communication and input/output processing in order to form the radio channel.

2.4.8. *Scope of the UML model*

The model proposed is flexible and open enough in order to enable the description of existing platforms as well as of operating systems and existing waveforms. The UML extension properties assure the capacity to describe the future hardware developments, waveforms or operating systems.

The next step for the SBC group is to standardize the element interfaces which are inserted into the presented architecture. The works defining the smart antennae and the digital IF started in 2005, the list of the societies that contributed and workplan having already been established. The objective of the standard for the digital part is to define the control interfaces, states, management and synchronization between the digital sender and receiver and the signal processing. The standard for smart antennae aims to define the commands and responses that make it possible to control such an antenna. For the two standards mentioned above, the description formats of the software and the abstractions will be defined.

The SBC group also created two *requests for proposal* (RFP) related on the one hand to the definition of the safety design and on the other hand to the implementation algorithms.

The works of the SBC group of the OMG clearly present a structured and mature approach of software radio promoted by the SDR Forum which wants to give value and durability to its research and development investments in open standards able to allow access to a number of inexpensive providers.

2.4.9. *The OMPT approach*

A similar approach appeared along with the Open Mobile Terminal Platform ("OMTP"). This initiative was created in June 2004 by eight international telecommunication operators. The purpose of the OMTP is to support the industry in the development of equipment whose capabilities would enable each mobile telephony operator to offer a competitive, unique and consistent service, irrespective of the user's equipment for the 2.5G networks and beyond. The OMTP is of the opinion that, to the greatest benefit of the subscribers, each access provider will use his own user interface, common and customized in its network for any type of equipment. This group will concentrate on the interface layer of operating systems and will support recommendations based on open standards and different technologies applicable in the mobile telephony industry. The OMTP wants to actively participate to the standardization process and to encourage the creation of an ecosystem of sellers and of hardware and software developers, according to its recommendations.

2.5. Conclusion

The entire approach presented in this chapter presents a structuration of industrial actors according to the present standards. It is not enough to establish a standard in order to be implemented. It has to be adopted by the manufacturers in order to create products and the customers must buy them. In the case of software radio, the mobile and basic station manufacturers are directly surrounded by questions in their design and organization model because the basic components are integrated by system operators in order to create the equipment. At the same time, there will appear manufacturers of components, architects capable of organizing and structuring them in order to meet the customer's needs, etc. The relations among different jobs and manufacturers will slightly evolve. The operator will be able to elaborate a terminal or a basic station adapted to his need by using a software radio architecture, which is internal to his organization or even independent, in order to create equipment that fits the customer's expectations as closely as possible.

The success of the approach depends on the appearance of industries which will design and market the components. For instance, beside some companies focused on the JTRS market, we can notice the appearance of products capable of becoming the backbone of the software radio architecture, like “eORB” of Vertel. The OSA/PARLAY or OMTP approaches encourage openness and flexibility in order to favor a more harmonious integration of the radiocommunications into our environment.

Chapter 3

Trade-offs for Building a Reconfigurable Radio Terminal

3.1. Introduction

If we go back along the history of the fixed or mobile terminal – the complete name being "edge communication terminal", which pertains to its belonging to a communication network – that we find today in millions of specimens and multiple forms, we find a casing in which communication functions are by analogy fulfilled, with a simplified microcontroller and an almost single function: the voice communication and a support of a rather wired unique transport network.

The first revolution was the digitization of networks and the digital communication era. The new edge terminal equipments are provided with analog-digital (AD) and digital-analog (DA) conversion modules and with modules of digital calculation. Thus, the hardware architecture of the terminal consists of a set of blocks of digital processing organized around a control module, the microcontroller, and of two "analog front-ends" representing the network interface and the interface toward the user (keyboard, microphone, etc.).

It is exactly this same structure that can be found in today's terminals, whether they are wired or wireless, with the difference that the access network has become

---

Chapter written by Marylin ARNDT, Eric BATUT, Jean-Philippe FASSINO, Florence GERMAIN, Tahar JARBOUI, Marc LACOSTE, Christian LEREAU, François MARX, Benoît MISCOPEIN and Jacques PULOU.

multiple and that many standards have appeared with universal and increasing needs for the required functionalities. Continuously in search of reducing the manufacturing costs, it seems obvious nowadays that the terminal will pass from a dedicated architecture to a more general and parameterizable one according to the needs. It is the same for the sub-block "radio interface". Facing the great diversity of 2.5G standards and the introduction of the third generation networks and new radio interfaces (WLAN, GPS, etc.), the future terminals must have the capability of dynamically adapting to the radio environment where they are located and this in all transparency for the user.

If we consider a multimode terminal, which is broadband and multichannel, the intuitive solution consisting of overlaying the different radio architectures leads to prohibitive cost, form factor and complexity, whereas the realization of a reconfigurable radio terminal represents a first stage toward the reduction of the costs expected in the long run.

This report leads us to imagine a much more open terminal where the software reconfiguration still makes it possible to use the most adequate radio standard (in terms of cost, quality of service, etc.). In this context, whereas the functionalities on mobile terminals do not stop growing, many technological challenges are to be raised to offer a powerful and certain reconfigurability, which will be an important factor of differentiation in a very competitive context.

The first parameter the reconfiguration will depend on is certainly the hardware architecture of the terminal. Firstly, the general architecture of a current terminal is coarsely described in order to extract the parameters which can be used by the reconfiguration action. In this section the various mechanisms of reconfiguration, which can be taken into consideration for the terminal, will be as well described, as well as the means of implementing them on a given architecture.

The second section will approach the hardware architecture of the terminal and in particular the flexibility brought by a reconfigurable terminal, especially for the reconfiguration of the algorithms of the baseband according to the state of the channel (intrastandard reconfiguration).

The third section will cover the software structure which is today an integral part of the terminal and whose importance will certainly not stop increasing. Certain crucial aspects of reconfiguration will be approached, particularly the compactness of the transmitted information, the safety of reconfiguration and the performances.

Finally, aiming to stress the open paths of research in this very wide field, the last part will tackle some open problems still unsolved today.

3.2. Architectures and reconfiguration mechanisms

In this section, we will show that the needs for reconfiguration naturally result in defining new architectures and mechanisms both at the hardware and software level.

3.2.1. *From scenario to architecture*

This section deals with the general problem of software reconfiguration. After a quick recall of the needs regarding software (re)configuration on radio terminal equipment, we will state some general principles pertaining to reconfiguration.

3.2.1.1. *Recall of needs*

This section briefly illustrates the needs concerning the terminal reconfiguration, by means of two typical scenarios:

(1) update of the remote terminal;

(2) adaptation of algorithms "to the execution".

The update of the terminal illustrated by the following scenarios implies the downloading of information (code, in general native, or data) with the aim of correction or improvement of the terminal (updated version):

– correction of bugs by distribution of software patch (for example, bug in the software of the camera, bug at the physical level);

– software update to improve functionality (for example, more powerful equalization algorithm, improving the communication in degraded channel conditions; control algorithm of the power saving the consumption of energy, change of audio/video codec);

– new radio standard;

– deployment of new services, or of new operational parameters.

Generally, these updates become effective only after rebooting the system. The updates can be developed by the equipment supplier, then validated and distributed to the users by the operator. Other models of distribution can be considered, according to the criticality of the updates.

The following scenarios illustrate the "online" modification of the terminal, i.e. the *adaptation to the execution* in order to optimize its performances according to the state of its resources or the change of environment (user requests, conditions of transmission, state of the network, geographical situation, mobility, etc.):

– update of entities of the protocol stack, following a request of the user asking for a particular service;

– optimization of the quality/resources compromise, in order to spare one of the terminal resources (batteries mainly);

– choice of a particular combination at the level of the physical layer, justified by the deterioration in quality of the radio link (situation of mobility, for example);

– adaptation of the algorithms of the radio layer (RRC + MAC) in order to optimize the coexistence of several users in a cell, following an evolution of the load network.

The adaptation operation is carried out in this case without rebooting the system and implies to a certain extent the continuity of service.

3.2.1.2. *General principles of reconfiguration*

(Re)configurability is a particular case of flexibility[1], when the latter refers to the architectural structure of the system considered.

The property of (re)configurability supposes the property of *modularity* and the *principle of assembling*. A system made up of interconnected modules will be configured by assembling modules and reconfigured by modification-replacement-addition-withdrawal of certain parts (modules or set of modules), with a minimum of interferences upon the rest of the system. The granularity of the modules is generally fine according to the technology (FPGA code, C code library, EJB component *enterprise Java bean*, etc.). Some of these modules are present during the execution, whereas others disappear, authorizing or not the dynamic reconfiguration during the execution.

---

1 Other forms of flexibility are respectively: the possibility of modifying the parameters, "specializable" character, or extensibility.

The potential actors of the reconfiguration are multiple: programmer, administrator, software entity, etc. The actor of the reconfiguration is not necessarily the same one according to the stages of the reconfiguration process (activation, decision, creation).

The typology of the reconfiguration modes is rather rich. Indeed, the action of configuration can intervene at various times in the life cycle of the system considered. We distinguish as follows:

– the static configuration intervening during the software deployment on the hardware platform;

– the semi-static (re)configuration intervening when booting or rebooting the system ;

– the semi-dynamic (re)configuration intervening in general on request, right before the execution. We also speak about "lazy" (re)configuration, when only the needs for the execution are taken into account;

– the dynamic reconfiguration appearing throughout the execution, thus enabling a "hot" adaptation of the code and data. In theory, the dynamic reconfiguration implies to a certain extent the continuity of service.

This section treats exclusively problems of reconfigurability (semi-static, semi-dynamic or dynamic).

Several stages characterize the reconfiguration. We can particularly distinguish:

– the activation, i.e. the detection of an event occurring in the terminal, or in its environment and/or the notification of a change. The change can be of different nature (lower availability of resources, increase in the applicative needs, deterioration of the quality of the radio link, etc.). It can be detected/notified by various actors (user, software sensor, hardware device, etc.) and at various moments (during the release of a service or during the execution);

– the decision consisting of determining the nature of the modifications to be caused: which entity has to be replaced? By which entity does it have to be replaced? Which process of replacement? When does the reconfiguration have to take place?

– the execution, which indicates the effective operation of reconfiguration. The realization can be based on various mechanisms, according to the need to guarantee or not the continuity of the service. This execution itself is split into several technical phases.

Lastly, the implementation of reconfiguration can be done either *by assembling* or *by composition*. The principle of design by assembling modules is already old and very widely used. A major difference between the main implementations of this principle is due to the fact that the assembling operator is explicit or not, via a model of programming. If an operator is explicit, we speak about *composition*. In this case, the composition operator is supported by entities of the same level of accessibility as the components themselves, thus supporting the structural handling of the systems by the program.

The impact of these general principles on the implemented architectures and mechanisms is detailed later on.

3.2.2. *Architecture and mechanisms for hardware reconfiguration*

This section presents the most promising analog and digital architectures in terms of flexibility, cost, performance and consumption for the reconfigurable radio.

3.2.2.1. *Functional architecture of a mobile terminal*

The partitioning in two blocks, namely a *front-end* and a baseband processing block, is the functional and traditional basic architecture of a mobile terminal. The role of the *front-end* is to filter and transpose the useful signal radio frequency (RF) in baseband and, reciprocally, other processings of the signal pertaining to the physical layer (equalization, encoding/decoding channel, etc.) and to the applicative layers are realized in baseband. This partitioning is illustrated by Figure 3.1.

In the current mobile terminals, the *front-end* is carried out in analog technology (AsGa/SiGe for the RF power amplifier, BiCMOS for the *transceiver*), whereas all the other specific baseband processings are carried out in digital technologies (CMOS). This technological fragmentation is necessary and answers the many performance/factor compromises of form/consumption/cost which are particularly constraining for embedded equipment. The partitioning of the analog and digital processings is closely related to the performances of the stages of digital conversion which are subjected to strong system constraints. Currently, taking into account the limited performances of the ADC (analog-to-digital converter), the digitization of the signal is carried out in baseband (see also Chapter 7).

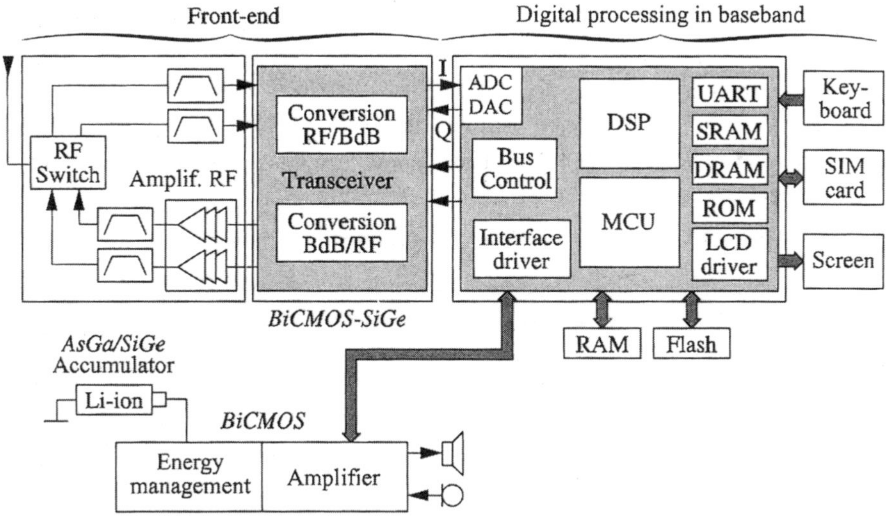

Figure 3.1. *Functional architecture of a mobile terminal*

The analog nature of the air interface imposes analog stages (RF power amplifier, RF filtering, etc.) which do not have the same flexibility for "programming/reconfigurability" as digital processing has and whose performances are specific to the concerned standard (dynamic of signal, bandwidth, etc.). Also, currently, for bi-mode terminals 2.5G/3G and even GSM bi-band, the tendency is the stacking of analog *front-end* architectures with narrowband. This approach cannot be possible for a reconfigurable terminal able to manage a great diversity of cellular standards and WLAN, in a context of strong competition. However, the duplication of certain analog components appears inevitable, particularly the RF power amplifier and the LNA (low noise amplifier).

3.2.2.2. *Reconfiguration of RF front-end, an outline on the new technological challenges and the candidate architectures*

There is no homogenity concerning the specifications of the various current standards of cellular type 2.5G/3G and WLAN. Also, one of the great challenges for the reconfigurable radio is the capability to manage a great diversity of specifications (bandwidth allocated, channel bandwidth, power of emission, modulation diagram, shape filtering, dynamics of power control, acceptable level of parasite emission, maximum level tolerated for the interfering signals, etc). In particular, the reconfigurable radio must necessarily have the possibility of functioning on various standards and thus of catering to a very broad spectrum. In practice, nowadays, if we consider the cellular systems 2G and 3G, WLAN networks, digital distribution networks, other systems (GPS, etc.), the spectrum considered should cover more than three octaves (DVB-T system, cellular systems 2.5G and 3G, 802.11a/BG, etc.). On the other hand, in order to cover the set of specifications relating to each standard, many technological challenges appear, among which we can mention:

– broadband spectrum and very high selectivity;

– diversity of modulation schemes (constant envelope modulation, non-constant envelope modulation) and their impact upon linearity and output;

– diversity of the RF output powers and dynamics of power control (90 dB max);

– diversity of multiple access methods and frequency duplex (impact on the RF head circuit and on the integration).

Among these many challenges, selecting the useful channel whose spectrum represents only one fraction of the band of the reconfigurable receiver appears to be vital.

For a narrowband receiver, this function is traditionally fulfilled at intermediate frequency (heterodyne assembly), even directly in baseband (homodyne assembly). Even if the heterodyne architecture makes it possible to distribute the selectivity, the gain and other functions, due to the multiplication of intermediate stages, it is dedicated to a given standard and thus it does not have the required flexibility for the reconfigurable radio. However, it should be noted that by distributing certain constraints along the chain, the specifications of the components, in particular on the ADC (*spurious free dynamic range* (SFDR), dynamics, bandwidth, etc.), are toned down.

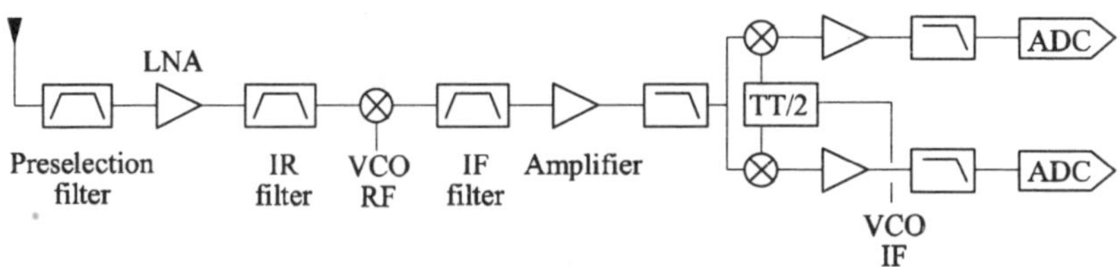

Figure 3.2. *Heterodyne architecture*

On the contrary, the homodyne assembly, which makes it possible to directly transpose the RF signal in baseband, offers a better integration, as well as a greater flexibility (the channel selection is carried out downstream from the ADC and is thus more easily programmable). However, this assembly imposes stronger system constraints (compensation DC *offset*, balance of I/Q paths in RF, IP2, etc.) and requires much better performances at the ADC level (dynamic, SFDR, etc.) than for the preceding assembly.

For a broadband receiver, the limited performances of the analog components would intuitively suggest that the architectures dedicated respectively to each mode are superimposed. This vision, which is antagonist to the requirements of embedded equipment (cost, form factor, complexity, etc.), is not acceptable. Thus, analog processings should be minimized, especially as digital processings offer a greater flexibility. It is thus natural and relevant planning to transfer the maximum of traditionally fulfilled analog functions (translation in frequency, filtering, channel selection, etc.) toward the digital one. For a broadband receiver, as minimal as the analog processing may be, the latter will have to process without distortion the strong power signals (adjacent channels, broadband signals) which can *a fortiori* interfere with the useful signal. This constraint has a considerable impact on the performances expected from the analog components (linearity, etc.) and in particular on the ADC converters (SFDR, jitter, etc.). The SFDR characteristic measured at the converter output defines the report of maximum tolerable powers (in dB) between strong and weak signals which could be simultaneously processed without distortion.

It is within this concept that the reconfigurable radio is perceived, the ideal approach consisting of digitizing the broadband signal closest to the antenna.

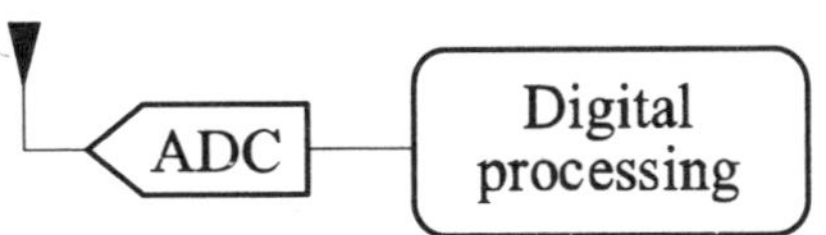

Figure 3.3. *Ideal reconfigurable radio*

This ideal approach would possibly require a filtering upstream and a broadband amplifier to bring back the signals and the noise to levels compatible with the characteristics of the converters. The latter, characterized by a very wide bandwidth, would have a subsequent sampling rate (more than 5 GHz), having as direct consequence a strong consumption which is incompatible with the embedded equipment.

In addition, according to this approach and by taking the GSM standard in the 900 MHz band, for which the specifications are amongst the most constraining, the ADC converter should present a minimum of 100 dB SFDR for a 9 dB CI, result which is beyond the ADC performances for a sampling rate of 1.8 GHz (see curve Figure 3.4 of [WAL 99]).

This characteristic curve confirms the interest in reducing the sampling rate in order to answer the new system constraints.

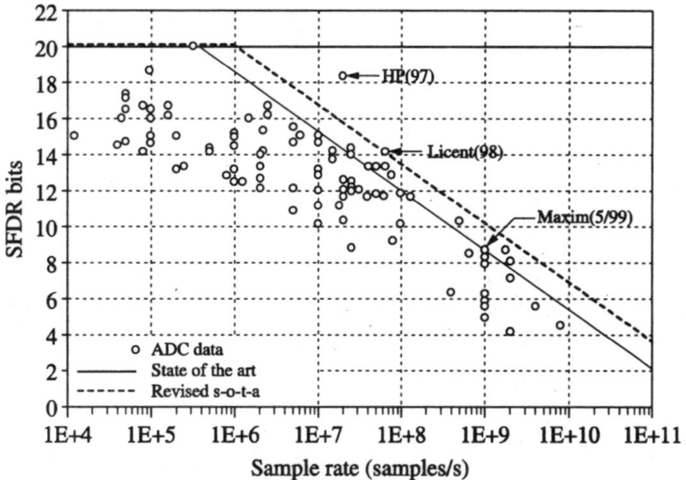

Figure 3.4. *ADC characteristic: SFDR versus sampling rate*

Various candidate architectures for the reconfigurable radio are presented in what follows. As we previously saw, in order to tone down the constraints on the digital conversion, it is advisable to transpose the useful spectrum (composed of various narrowband channels) into a lower intermediate frequency. This operation can be carried out either by analog transposition in frequency, or by a sampling of the preselected spectrum. These two concepts are currently the object of many studies and publications.

The first technique consists of digitizing the signal at IF (intermediate frequency). It refers to a broadband architecture with double or simple IF using the technique of subsampling (or undersampling) at IF. After filtering and amplification (LNA), the useful spectrum is relocated at a low intermediate frequency (compatible with the ADC performances) via one or several analog stages. The digitization of the useful spectrum, which is realized at IF and according to the subsampling technique, makes it possible to transpose digitally with weak sampling rates the useful spectrum to a lower IF before selecting the channel. In fact, the observation of certain constraints makes it possible not to literally respect the traditional criterion of sampling.

However, the subsampling technique inevitably imposes many technological compromises (factor of quality, SFDR, sampling rate, etc.).

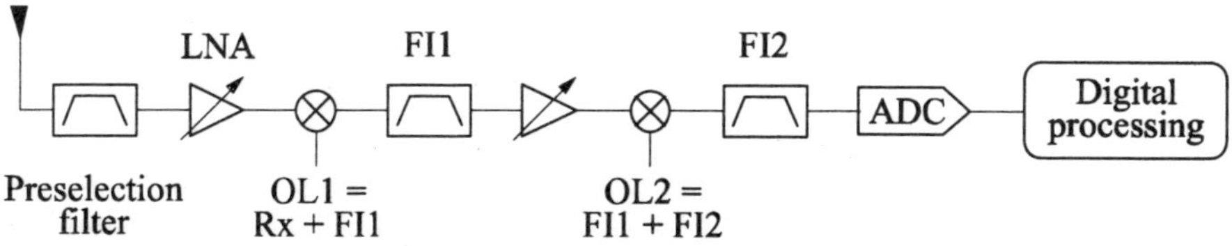

Figure 3.5. *Digitization of the signal at IF, technique of subsampling*

The second technique consists of a direct sampling of the RF signal followed by a digitization at a low intermediate frequency. For this approach, after preselection and amplification filtering, a judicious and direct subsampling of the spectrum is carried out, an operation which makes it possible to relocate the RF spectrum at a lower intermediate frequency. Then, by the decimation technique, this intermediate frequency can be relocated up to an intermediate lower frequency and compatible with the performances of the ADC converters before the digitization of the useful signal. This innovative architecture was tested for Bluetooth [MUH 04] and could be possibly conceivable for the reconfigurable radio.

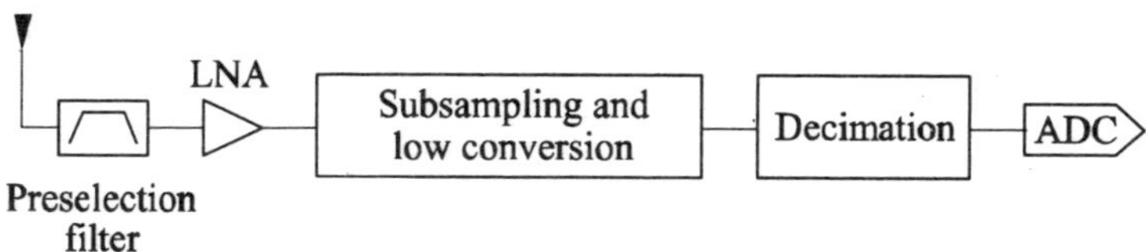

Figure 3.6. *Sampling in RF and digitization at IF*

3.2.2.3. *Digital reconfigurable architecture*

The problem of the digital architecture of a terminal is rather different from the analog part, insofar as the choice of the material resources is made only after a functional study of the data processing sequence to implement, during which the algorithms of signal processing were evaluated in performances and load of necessary calculation. For the implementation of these data processing sequences there are two main tendencies: a choice directed toward the specifically designed *hardware* components, which are managed by a microcontroller, and another one which places a core of the signal processor (DSP) in the center of the digital architecture. The two sections below briefly present these two architectural options.

The reconfigurability of the baseband chain processing can be obtained by two great types of mechanisms. The first one consists of operating a transformation of the data path in order to make active the treatments blocks already implemented on the hardware resources. This option enables a fast reconfiguration but it is not very satisfactory in terms of scalability, neither of the terminal's functionalities nor for the optimization of the hardware resources because it is necessary that the latter is able to implement all the treatments of all the supported modes. This solution is rather adapted to the dominant hardware architecture. The other solution is to reconfigure the treatments by reprogramming the components of the architecture, which is completely adapted to the other architectural solution, based on the core of processors. From the point of view of performance, this type of solution has the advantage of flexibility, at the cost of the speed to be reconfigured.

For the reconfigurable component architecture, the set of the processings in baseband is mainly articulated around the microprocessor and a reconfigurable hardware subsystem which is charged to carry out the various processings necessary to the radio modes.

The intensive processings are dealt with by reconfigurable *hardware* blocks which carry out the head digital processings, the processings of channel (de)coding and other operations on useful data (multiplexing of channels, interleavings, etc.). These blocks, which are, for example, parameterizable FPGA or ASIC, offer a good reconfiguration capability and make it possible to ensure a reconfigurability between various radio standards.

The role of the microprocessor is:

– to configure the hardware components on which the physical layer of the radio is executed;

– to carry out the protocol layers MAC and RLC/RRC of the standard, known also as high layers.

All the intensive calculation parts are dealt with by the reconfigurable *hardware* circuits, which makes it possible to limit the consumption of the terminal, whereas the microprocessor should manage only events whose frequency of occurrence is close to the slot or frame period.

The hardware architecture centered on a DSP is basically different from the preceding one even if it seems to relate to only one weak part of it. In fact, the reconfigurable *hardware* components are replaced by components of much lower granularity since they carry out only one task: interleaving, channel coding or *rake*, for example. They can be related with material accelerators of the DSP, even offset material operators (also called coprocessors).

In addition, we add to these coprocessors a powerful DSP. A microprocessor is always present and ensures the processing of the protocolar stack above the physical layer. In this diagram, the DSP is the central body of calculation of the physical layer because it carries out part of the treatments, as well as the source coders, but it is also the component which configures the whole of the operation parameters of the physical layer and ensures the sequencing. Therefore, all the intensive calculation functions as well as those which work at the chips rhythm ask for a number of instructions per second insofar as they cannot be implemented on the DSP but rather on material components. The latter receive the data to be processed and then restore them to the DSP.

3.2.2.4. *Comparison*

The comparison of these two possibilities of reconfigurable architectures of the digital part can be summarized by opposing two major criteria:

– the hardware type architecture is more interesting in terms of consumption of energy because it is founded on dedicated coprocessors and on an MCU, which are sequences typically at clock speeds lower than 50 MHz, whereas the architecture centered on a DSP requires at least speeds of about 250 MHz;

– the architecture centered on a DSP is on the other hand much more advantageous considering the flexibility and evolutionarity aspect of the terminal and moreover it requires less design time than the component-based architecture.

3.2.3. *Mechanisms for software reconfiguration*

This section covers the problem of software reconfiguration by assembly initially and then by composition.

3.2.3.1. *A first stage toward reconfiguration: assembly designing of extensible systems*

The first response regarding the report of rigidity of traditional systems is the concept of extensible system. The objective of this approach is to enable the applications to personalize the stock management by addition/replacement of system abstractions. Flexibility relies on the "modularization" of strategies in the form of libraries or modules, which are charged under dynamic extensions [SIL 98]. This is a weak form of reconfigurability, implemented by code assembly and primarily directed toward the dynamic addition of modules favoring the execution requirements. The modifications resulting from an extension can be loaded inside the core (intracore extensions) or at the applicative level (extracore extensions). The consequences in terms of safety in particular are covered in section 3.4.2.

The mechanism of dynamic loading is justified by the willingness to limit the size of the executables while avoiding loading code parts which will never be activated. The principle is as follows: all the routines are presented on a permanent support (disc or memory flash) in a relocatable binary format. The main program is memory loaded and executed. When a routine is invoked without being loaded, a dynamic link editor is called. It loads the routine in the execution memory and updates the references at this address in the program. The principle of dynamic connection is similar to the principle of dynamic loading. It is especially used for the libraries system. A *stub* is included in the image memory for each reference to an external routine. The *stub* indicates how to locate the routine in question in the memory, or how to load it if it is not present. The *stub* is replaced by the address of the routine at the time of the first call to the routine. This mechanism can be extended to the case of updating bookshops, enabling all the programs which refer to a bookshop to always use the latest version (by simple intervention on the *stub*).

The update of the references carried out at the time of a dynamic link editing authorizes, in a tiresome way, the capture of dependencies, and thus the replacement of a (set of) module(s) by another. However, the reconfiguration of a system supposes a lot more than only the management of dependencies, in particular the identification of a state suitable to support the reconfiguration operation (named "stable state" from now on), the atomicity of the operation, the recovery of the released resources, etc. With the approach by dynamic link of modules, none of these operations is possible without specific tools or programming convention.

We will see in the following section how the "components" approach, in what it explicitly expresses the architectural entities of a system, facilitates the effective implementation of the reconfiguration.

3.2.3.2. *A second stage toward reconfiguration: the compositional approach*

Object-oriented programming (see Chapter 2) is one of the most reduced and most completed attempts used in order to design a system by assembly of reusable software entities dynamically created. However, the progress of the object model in terms of reuseability (encapsulation, heritage, polymorphism) contributes only partially to obtaining a structuring architectural vision, disclosing the concept of dependence between the entities.

The limitations of the object approach motivated the approach called "component-based approach" [CZY 02, HEI 01, LEA 00]. A component is at the beginning a unit of rollout and software structure, specified independently of its implementation. A component presents, as well as an object, interfaces known as "functional", translating the services which it offers ("server" interface), but also – and this is one of the differences from the object model – the services which it requires ("customer" interface). The components are assembled by means of an operator, called "link operator". A link connects two interfaces, customer and server respectively. A fundamental characteristic of the component models is to make explicit (and thus programmable) these links by integrating them in the programming model. Working by link program, as well as by the component one, is a key-element of reconfiguration. A second characteristic of the component-based models is to authorize the capture of the compositional hierarchy [BRU]. A system of hierarchical components is then described in the form of an overlap of "primitive" components within the "composite" components. Just as for the links, such a system authorizes the program handling of its hierarchical structure.

A system intended to be dynamically reconfigurable must have access, during its implementation, to its own structure ("meta-information" on the system). Let us consider, for example, the operation of replacing a component by another. This operation will bring into play (in order to optimize) the following "elementary" operations:

– the disconnection of the component to be replaced (and thus its shut down in a state favorable to support the reconfiguration);

– the possible recovery of the resources used by the replaced component;

– the modification of the links incidental to the component to be replaced, in order to connect them to the new component;

– the instantiation of the replacing component (for example, starting from data: code, parameters, etc, transmitted by a wired or radio link);

– the connection of the new component (state transfer if necessary between the old and the new component, etc. which supposes that the latter was preserved or that its status was stored);

– the activation of the new component.

· In order to implement the operations above, it is necessary to dynamically have access (in the sense of observation/modification) to a certain number of architectural characteristics: topology of the system, status of connections, etc. This refers to one of the facets of what we call "reflection". Reflection indicates the capacity of a system to interact with its own definition, in terms of its structure or behavior. In the first case, we speak about "structural reflection" and in the second, of "behavioral reflection". Reflection is in general supported by a "meta-level", implementing the functions of observation and modification (introspection/intercession) of the basic level via a meta-model of the latter [PAT 87].

The component technologies offer a unified formalism providing at the same time the expression of basic level and meta-level, as well as the expression of the interaction between one and the other (implementation of the *meta-object protocols*, MOP) [KIC 91].

3.3. Compromise for the hardware reconfiguration

3.3.1. *Baseband: to benefit from the reconfigurability in order to limit consumption*

This section is dedicated to the use of reconfigurability in order to optimize the complexity of the signal processing algorithms in the physical layer of the radio. In fact, even if at first the software-defined radio seems greedier in energy, it also opens new possibilities. The reconfiguration of signal processing blocks makes it possible to choose the most adapted algorithm according to the channel conditions or to the required service. The dynamic reconfiguration of the algorithms decreases the number of operations carried out by the arithmetic unit meant to decode the radio signal and thus to save the terminal battery. Through the example of equalization and decoding channel, the compromise between flexibility and consumption /complexity will be presented below.

3.3.1.1. *Equalizer*

Under the name of equalizer, we consider the processing blocks necessary for the compensation of the distortions of the received signal, following its passage in the radio channel. The main component is, coarsely, a filter adapted to the channel, but it is advisable to also take into account the channel estimation block, making it possible to determine the parameters of the propagation channel (multi-path intensity profile, i.e. the number of multiple paths with their delays and respective complex attenuations). Indeed, this task operates on the oversampled date. As a result, its impact on the design load is very important. As an example, an evaluation of the complexity of the equalizer for a UMTS FDD receiver shows that for the majority of possible services, the channel estimate requires more operations to be carried out than the *rake* receiver (see Chapter 5 for details on this subject). The numbers in the table below come from a calculation of the operations necessary to support the various UMTS/FDD throughputs, by making the assumption not very realistic but enabling the comparisons, of their implementation on a DSP which can parallel 4 ALU. They are expressed in MHz, in order to symbolize the load of calculation of the target DSP.

MHz versus throughput (kbps)	**12.2**	**64**	**144**	**384**	**2 Mbps**
Channel estimation	120	125	127	127	66
Rake	35	40	41	125	290
Total (MHz)	155	165	168	252	356

Table 3.1. *Estimate of complexity of the UMTS FDD equalizer*

The sensitive variations of one service to the other come mainly from the service and the corresponding parameters of the communication as the variable number of the control *chips* present in the *slot* or the diagram of code attribution to a user.

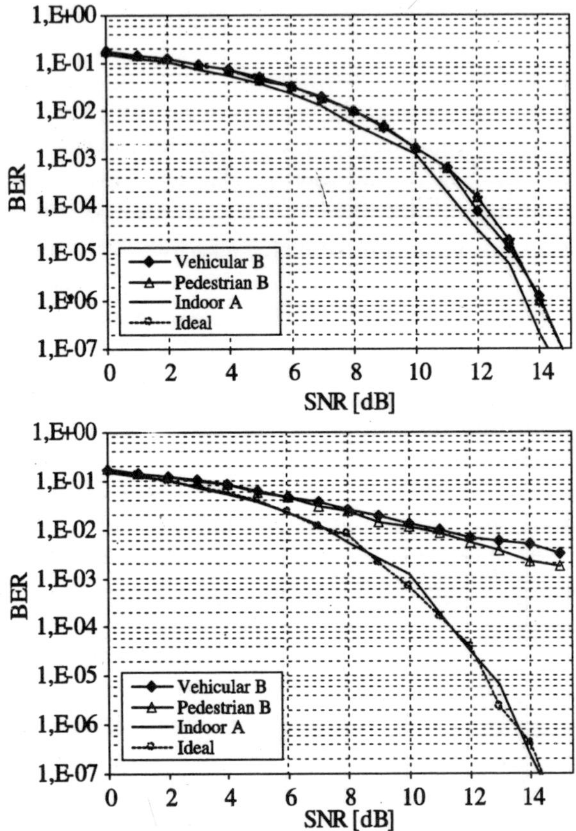

Figure 3.7. *Impact of equalization in FDD for a short code*

Traditionally, the FDD equalization diagram is often reduced to the *rake* because this mode of the UMTS is associated with important spreading factors, which guarantee a long symbol time in front of the average spreading of the propagation channels. An equalizer, in charge with processing the intersymbol interference after the *rake*, is sometimes considered but often omitted in the chains of reception usually described in other works. However, it seems delicate to neglect its impact in extreme cases of using short codes which can generate very detrimental intersymbol interferences for the receiver as Figure 3.7 illustrates.

We see on the bottom graph that the equalizer's absence causes a degradation of the performances because for a too short spreading code (here 4 *chips*), the symbol time can prove too short with respect to the spreading of the channel. The generated intersymbol interferences cause a noise floor, as seen in the bit error rate curve. We thus see that the presence of an equalizer behind the *rake* is not always necessary but, in certain cases, can prove to be necessary to guarantee the quality of the radio link.

However, we also notice in this example that the equalization block [CIB 04], if it can acquire a sufficient flexibility, can make it possible to implement processings adapted to propagation conditions. A minimal level of performance can be thus guaranteed while optimizing the design load factor. Now let us see other potentially reconfigurable parameters related to the channel estimation and the *rake* receiver.

For the channel estimation, there are many sources of flexibility, due to which it is possible to strongly reduce the calculation complexity when the radio environment can do it. We can mainly mention the following cases:

– the *slots* rate for which the channel estimation is partially made. Indeed, rather than making a complete channel estimation for all the *slots*, it is possible to refresh the estimation from one *slot* to another, by a simple correlation on only one time *chip*, which strongly decreases complexity. Thus, the complexity estimation in Table 3.1 is made for a 66% rate; if we fix it at 87% (i.e. 13 *slots* in the frame), the load of the channel estimation passes, for 384 Mbps, from 61 to 27 MHz;

– the depth of the research window of secondary paths: before making the channel estimation, it is necessary to fix this parameter which describes the maximum delay of the last path supported by the equalizer. In the case of channel estimation by correlation, the depth of research is the number of correlation points to be calculated. Typically, in the case of the *indoor* situations where the radio channel is short, the channel estimator has thus a great interest in working on a reduced research window. Always in the case of Table 3.1, the estimate is made for a 40 *chip* window, that is to say 10 µs. By reducing the window to 2.6 µs, the load of the channel estimator is then of 23 MHz, that is to say 62% of complexity reduction;

– the oversampling factor upon which the correlation complexity depends directly. Thus, in a case of stable channel for which the paths are perfectly discernible, the division by 2 of the oversampling factor causes a reduction of 50% of the correlation complexity of the channel estimation.

For the *rake*, the main source of flexibility comes from the number of fingers of the receiver which we implement. Each finger is necessary to recombine a path of propagation and thus to recover a part of the energy transmitted by the transmitter. The channels described in the UMTS standard are recorded to some extent in Table

3.2; we note that the majority of energy is located in the few first paths, even if 6 paths are considered in general.

		Path 1	Path 2	Path 3	Path 4	Path 5	Path 6
Indoor		0 dB	– 3 dB	– 10 dB	– 18 dB	– 26 dB	– 32 dB
		0 ns	50 ns	110 ns	170 ns	290 ns	310 ns
Pedestrian		0 dB	– 9.7 dB	– 19.2 dB	– 22.8 dB	/	/
		0 ns	110 ns	190 ns	410 ns	/	/
Vehicular		– 2.5 dB	0 dB	– 12.8 dB	– 10 dB	– 25.2 dB	– 16 dB
		0 ns	300 ns	8,900 ns	12,900 ns	17,100 ns	20,000 ns

Table 3.2. *Indoor, pedestrian and vehicular A channels of 3GPP*

In the light of the information contained in Table 3.2, we can significantly reduce the *rake* complexity by reducing the number of the receiver's fingers. At the extreme, one can very well consider that the terminal can deliberately reduce this parameter to the minimum (for example, 2) if the measurement of the signal-to-noise ratio at output is sufficient or if a strategy for reducing energy is required by the reconfigurability manager. Returning to the figures of complexity presented above, we note that in the case of an *indoor* channel, the recovery of the first two paths (guaranteeing a recovery of more than 80% of the energy of the transmitted signal) makes it possible to reduce the complexity of the *rake* by 50%.

For the equalizer, it was previously shown that its activation was not always necessary. If it is necessary, it is possible to make a compromise between complexity and performance by reducing the depth of the equalizer, that is to say the number of equalizer coefficients to be calculated.

3.3.1.2. *Channel coding*

The objective of channel encoding is to reduce the probability of errors on the transmitted information. In the case of convolutional codes, the best known and mostly used decoding algorithm is the Viterbi algorithm which provides the maximum of probability on a code word. The principal disadvantage of the Viterbi decoder is its complexity which grows exponentially with the constraint length of the code. In the case of correcting codes with a significant constraint length (for the UMTS or the IS-95 the constraint length is 9), the number of operations becomes very high. The complexity of sequential decoding is independent of the constraint length of the code and, under good conditions of the signal-to-noise ratio, sequential decoding thus requires fewer operations than the Viterbi decoding.

The sequential decoding MP (maximum of probability) is largely presented in articles [EKR 94, HAN 02, FEL 02]. The general idea of the algorithm is to develop the node of the lattice which will probably lead to the optimal way and to eliminate the nodes which have only suboptimal descendants. This decoding algorithm provides exactly the same performances in terms of binary error rates as the Viterbi decoding because the two algorithms minimize the same metric. On the other hand, it should be stressed that the complexity of the MP sequential decoding is a random variable, which leads us to consider an average complexity contrary to the Viterbi algorithm where complexity is constant.

Figure 3.8 illustrates the average number of metrics calculated using the Viterbi algorithm and the sequential decoding. We note that for the good SNRs the sequential decoding algorithm calculates a number of metrics far lower than the Viterbi algorithm. For very good SNRs, the number of metrics calculated by the MP sequential algorithm is even optimal, i.e. 2.

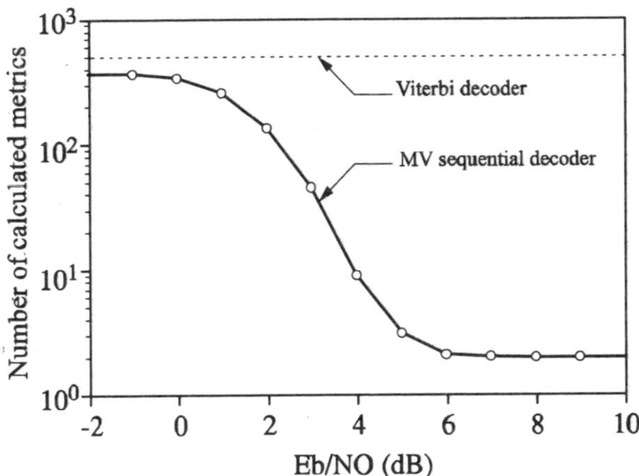

Figure 3.8. *Numbers of metrics calculated by symbol of information for the Viterbi decoding and the MP sequential decoding of the convolutional UMTS code of constraint length 9 and generating polynomials (561,753) in the case of a Gaussian channel according to the signal-to-noise ratio*

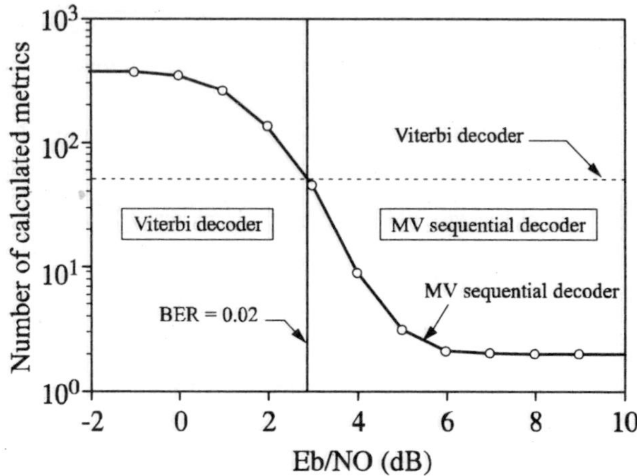

Figure 3.9. *Numbers of operations per symbol of information for the Viterbi decoding and the MP sequential decoding of the convolutional UMTS code of constraint length 9 and generating polynomials (561,753) in the case of a Gaussian channel according to the signal-to-noise ratio*

However, the calculation quantity is not given solely by the number of traversed branches but also by the cost of expanding a branch of the lattice (research and insertion in the stack) which is higher for the sequential decoding with respect to the Viterbi decoding. It is therefore interesting to know in which zone the Viterbi decoding is most interesting and vice versa. This problem clashes with the evaluation of the complexity of expanding a branch which depends largely on the selected implementation. A rough estimation provides a complexity 10 times higher for the sequential decoding algorithm. Figure 3.9 presents two SNR zones: in the first one the Viterbi algorithm is the least complex and most capable to decrease the load of the machine; in the second zone, which covers most of the useful SNR (BER $<10^{-2}$), the sequential decoding algorithm is the most powerful.

According to the quality of the channel (SNR, for example), the radio is automatically reconfigured and selects the algorithm requiring fewer operations, the two algorithms offering exactly the same performances in terms of binary error rates. The example of convolutional encoding shows that a profit of two complexity orders of magnitude can be reached by best choosing the decoding algorithm (the Viterbi decoding or sequential decoding). In the same way for iterative decoding (turbocode, for example), reducing the iterations' number at the time of decoding linearly decreases complexity.

3.3.1.3. *Conclusion*

The reconfigurable radio brings flexibility in the choice of algorithms while making it possible, for example, to choose the most relevant algorithm according to the channel conditions. For the channel estimation, the equalization and the decoding channel, we see that the adjustment of certain significant parameters can only have a strong impact on the consumption of these blocks, but this reconfigurability is made within a framework of total control of the performances, in order to avoid a drift in the calculation process which would call into question the integrity of the radio link. Generally, it seems delicate to resort to such mechanisms in situations of difficult channel or strong mobility. Therefore, for a favorable environment usage, it is perfectly possible to practice such processing simplifications, with a progressive management of the reduction of complexity (and thus of the degradation of the equalized signal) in order to converge toward the best compromise between the performance of the radio link and the consumption of the receiver.

3.3.2. *Mechanisms of reconfiguration and control: flexibility versus genericity*

The implementation of a reconfigurable multistandard physical layer within a mobile terminal poses a certain number of problems by which the terminals known as monostandard are absolutely not concerned. The conceptual border is blurred between *data* and *control*, which are two fields considered before as absolutely disjoined, the functional cutting of the emission and reception chains is called into question and the right level of exposure of the material to software is being studied again.

3.3.2.1. *Is the absolute separation data/control always accepted?*

A traditional data processing sequence establishes a rather strict separation between data and control. Control consists of the code being carried out on the arithmetic unit concerned (MCU or DSP), code stored in RAM or possibly in ROM for the invariant functions and the *bootstrap* code. As for them, the data are *exclusively* transmitted or received through the radio link and consumed by the applicative layers. This simplistic diagram is valid for the communication standards defining in an exhaustive way all the possible manners to transport formatted data. The data do not have any influence on control and only pass through the low layers before arriving within range of the user by means of dedicated applications.

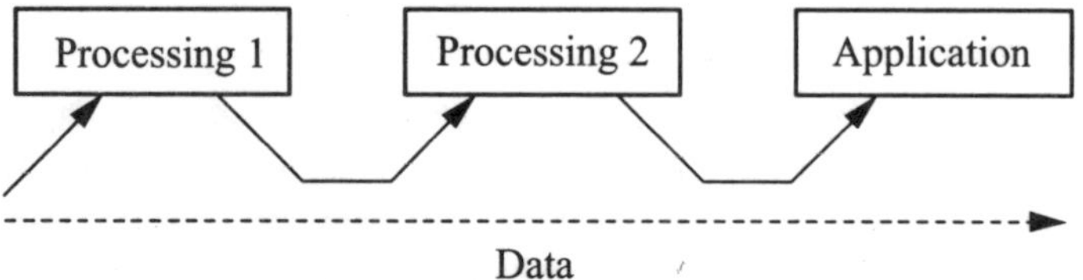

Figure 3.10. *One-way data traffic*

The situation is quite different for standards offering more flexibility in terms of the various possible transport formats. For a standard like the UMTS, for example, a first decoding of data or part of the transmitted data will make it possible to choose the way in which these same data will be decoded. Be it by TFCI (*transport format combination indicator*), when it is used, or by means of *transport format* when it is defined in an unambiguous way by the high layers, a part of the transmitted data will directly influence the code which will be carried out to decode the other part of it.

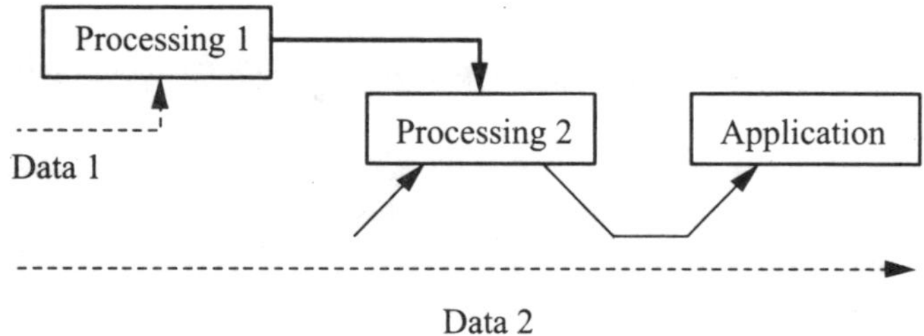

Figure 3.11. *First loop data – control*

Finally, adopting a reconfigurable physical layer comes again to disturb the data flow within the control, practically destroying the separation which could have been maintained before between the two of them. In the preceding example, the processings to be carried out are known to the receiver, who will be satisfied to choose one processing rather than another before applying it to the rest of the data. A reconfigurable system will enable the processing to be directly instantiated from the received data, enabling the calculation to carry out a certain representation of the received *data*, thus promoting them to the rank of control code. This example occurs, for example, during the downloading of a new channel decoder through the radio link, or of a new equalizer.

The following emergence of really reconfigurable systems, including in terms of their radio interfaces, will thus involve a thinning of the border nowadays maintained between data and the control code, an incursion of data within the control code which can occur in certain cases. It goes without saying that many problems of safety and software infrastructure will then arise. These points are approached in the following sections.

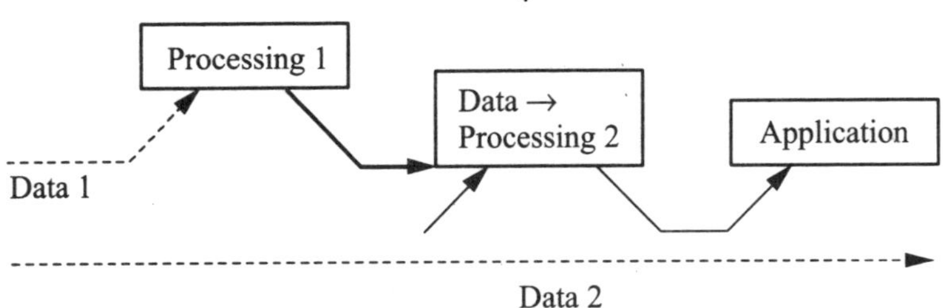

Figure 3.12. *Data promotion in control code*

3.3.2.2. *Is a fixed functional clipping of the processing still relevant?*

The reception chains implemented in the terminal equipment in conformity with the current standards can all be split in a certain number of stages relatively well identified, with clear and precise goals and interfaces. We can mention, for example, the formatting, the channel estimation, the equalization, various possible channel decodings, as well as a certain number of blocks doing nothing but handling the data without modifying them by calculation before presenting them to the consumer in adequate higher layer format.

Once the interfaces are defined very clearly, we can imagine being able to download a new equalizer while using the transparency data/control exposed to the preceding point. The new equalizer having the same inputs and outputs, we imagine being able to easily carry out a replacement, even temporary, of the equalizer already present in the data processing sequence by the equalizer which we has just been downloaded through the radio link. It should be noted, however, that the details of the implementation of such a replacement are not given here. For instance, in the replacement of a complex equalizer by a simpler and faster equalizer, many problems can arise, such as bandwidth requirements at the input/output of the equalizer, power consumption, etc. This is obviously even more an issue in the case of hardware change! It is therefore recommended to validate *a priori* the change of any software/hardware component before releasing the updated software/hardware component. New certification procedures may be defined in the context of reconfigurable devices.

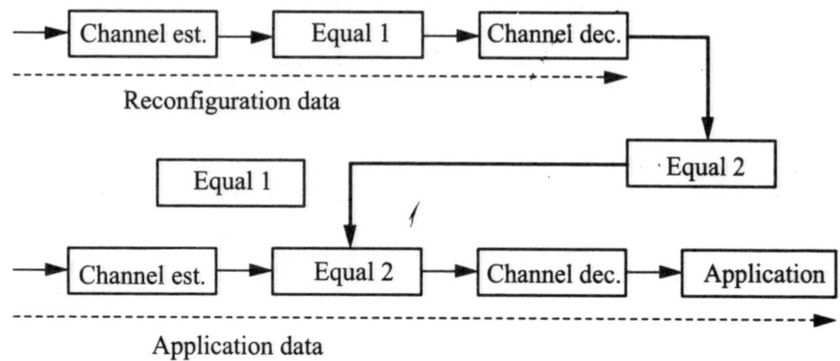

Figure 3.13. *Replacement of an operator within the processing sequence*

This functional split has the unquestionable advantage to standardize the roles of each block and to define very precisely their interfaces, which is an indispensable condition for a profitable interconnection to take place. It has, however, the disadvantage of freezing these same interfaces and of rigidifying the structure of the processing sequence itself, which can thus lose a degree of freedom. What happens when a reconfiguration event involves the replacement of two disjoined blocks, contiguous or not, by only one block providing the function of the two blocks replaced by it? If we take the example of turbo-equalization, which is the iterative process of equalization based on the symbols resulting from channel decoding to adjust its decisions and to improve the quality of its outputs, how can we include the concept of interblock iteration in the new processing sequence?

The difficulty which arises then is that of the guarantee of the integrity of the processing sequence: without a strict control of the interfaces removed or reintroduced during the replacements of operators by download, the total coherence of the chain becomes delicate to maintain. Once two operators are replaced by one sole third one, as in the turbo-equalization example, how can we manage the case of data requiring another channel decoder? Is it necessary to download again a standard equalizer and an adequate channel decoder or is it enough to add the adequate channel decoder by "short-circuiting" a part of the turbo-equalizer?

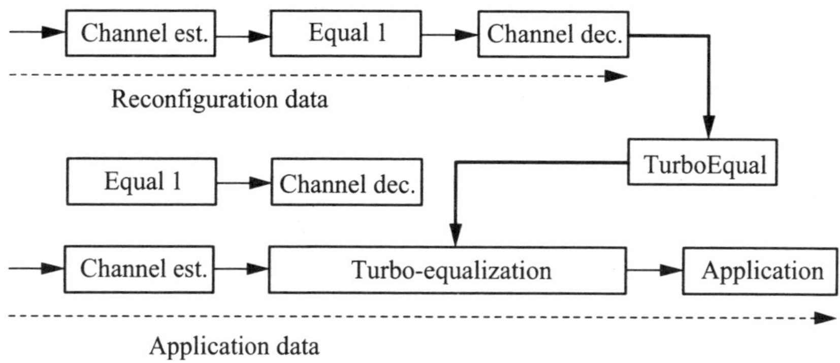

Figure 3.14. *Downloading an operator including several operators of the original chain*

3.3.2.3. *Which degree of exposure of the hardware to the reconfiguration mechanism?*

The possible modification of the implementation environment of the processing chain by the data resulting from this same processing chain necessarily implies a certain degree of exposure of the material with respect to the authority controlling the reconfiguration process. The term "hardware" is used here in the broad sense to indicate all hardware components, programmable or not, which ensure the implantation of the various algorithms which can in one way or another be reconfigured.

The control layer, be it hidden within the processing layer or orthogonal, must be able to modify the software or hardware configuration of various chain operators at a moment when they are inactive, so as not to disturb the chain operation. Planning problems being put aside, it is necessary to find a way to expose this hardware configuration optimizing the reconfiguration process in its entirety. The right balance must be between a very detailed exposure of the least architectural detail of the terminal, which in addition to having effects on the safety of the implementation environment can make the description of a reconfiguration order particularly tiresome to work out and bulky to transmit. A potential way forward could be to define a limited set of reconfiguration possibilities; the process of reconfiguring the terminal would consist in that case of simply indicating the selected possibility.

3.4. Compromise for software reconfiguration

3.4.1. *Reconfigurability and compactness of transmitted information*

The reconfiguration of the radio links involves the (new) code loading on the terminal. If this loading is carried out by the radio link, certain considered scenarios (for example, the distribution of an update patch to all customers of an operator) show situations leading to a Hertzian distribution toward a great number of terminals. The saturation of the radio link is thus not to be excluded, especially if the studied scenario lies within the framework of a crisis situation in which the limits of the transmission capacities are already reached. From this point of view, the size of the transported code can prove to be an important constraint. However, this criterion interferes with others, such as portability, safety, size of the support of execution and performances, which leads to a compromise. Between the "source" form – used during the development – and the machine language, a program passes through a series of translations which will transform this program from one form into another. One of these forms will be used for the transport of a downloadable code. This section provides some background information of the aforementioned compromise according to the selected form.

A priori a language known as "high level" makes it possible to express a program in a way more compact than the machine language. It is the same for the language specialization, which aims at providing to the programmer/specificator entities corresponding to the "trade" objects and their relations. However, this intuition is counterbalanced by the need for keeping a form easily readable by the programmer, which involves, for example, the use of reference symbols (identifiers) whose impact on the code size is significant. The generation of an intermediate code obtained by replacing the internal identifiers with less greedy symbols (for example numbers) makes it possible to reduce the size of the code. This processing is used with other techniques in the *byte-code* generation from a Java source file [JAV] but other languages, which are less widespread, also use an intermediate form (see OCAML [OCA]).

Taking into consideration only the size criterion, the choice of a *byte-code* intermediate code seems to gain ground. Some experiments undertaken on the Java compiler toward native GCJ [GCJ] indicate that the size of the native code is partly due to the obligation to introduce sequences (of native code) intended to implement the verifications that can guarantee the properties of safety (*type-safe*) brought by Java (verifications typically carried out by the couple auditor/JVM in an interpreted version). In order to maintain these properties, the presence of these sequences in a downloaded native code will have to be guaranteed by a mechanism of certificate or identification of the code origin attesting the way in which it was generated. However, it is possible to remove these sequences, at least partially, and thus to reduce considerably the size of the transmitted code so that certain verifications can be entrusted to hardware devices available on the used processor (memory protection devices of MMU type on ARM architecture)[2].

This native code *versus byte-code* discussion must also take into account the size of the target infrastructure and the speed performances. If the *byte-code* infrastructure is *a priori* bulkier than the "native" infrastructure, this advantage must be counterbalanced by the relative importance of the "libraries" necessarily embedded into the two solutions, which makes us go back to the preceding discussion. Under the angle of performances, the solutions of embedded compilation "on the fly" (seldom available on small targets), i.e. "before implementation" like that proposed by the company Esmertec [ESM], can compete on the plan of performances with the "native" solutions… at the expense of the infrastructure's size.

3.4.2. *Reconfigurability and safety*

Protecting the resources of a mobile terminal is critical for a telecommunications operator. It requires strong measures of safety which must be located in the lowest software layers like the operating system in order to serve as a base for the protection of the system. The changing execution environment of a mobile terminal comprising multiple types of networks with crossing of several logical barriers can endanger this safety. That calls into question even the reconfigurability of the terminal which has recently extended to the operating system [DEN 02]: it is from now on possible to download potentially hostile software components.

An effective control of the access to the resources of the terminal supposes a clear separation between the *safety policy*, which defines the rules conditioning the access, and the *mechanisms* meant to enforce this policy. The latter constitute the *confidence basis* or TCB (*trusted computing base*) gathering all the vital components for the safety of the system [USD 85]. They must guarantee the following properties:

– *safety*: no illegitimate access to the resources, no bypass of the protection mechanisms (*complete mediation*) whose integrity must be assured, no abusive propagation of the administrator's rights of access (*minor privilege*);

2 The issue in this case is the necessary sophistication of the hardware support.

– a minimal impact on the *performances*;

– *flexibility*:

- support of several safety policies,

- access control to variable granularity,

- dynamic management of the rights of access and of their delegation;

– *simplicity* of design, use and administration;

– *confidence*: simplicity and reduced size of TCB which must be able to be certified as correct by a third party of confidence;

– it is difficult to find a right balance between these properties, which are often contradictory. However, several design parameters make it possible to arrive at a compromise: we will mention here:

- the type of core considered,

- the underlying safety model,

- the localization of the protection mechanism, and

- the level of assurance measuring the degree of confidence which can be granted to the system.

3.4.2.1. *Type of kernel*

The operating system offers to the applications an abstract sight of the resources of the terminal and manages their division and protection. It is generally composed of:

– a *kernel* having access to all hardware resources where the processor can carry out the favored instructions in a supervisory mode;

– the highest level *service system* used by the applications.

The architecture's safety strongly varies according to whether the services belong to the field of protection of the kernel or whether they are separate processes. Creating distinct fields favors safety: each application remains confined in its space of addressing and cannot corrupt the rest of the system. However, this separation penalizes performances: each request of access to the hardware coming from the applicative field of protection requires an expensive call system because of context changes with the field of the kernel protection.

The *monolithic kernels* like Linux or PalmOS integrate the services in the field of protection of the kernel. The system obtained is without an apparent structure, of large size, and not very adaptable. Due to its complexity, the system contains many safety faults: the attacks remain difficult to confine and often cause the crash of the whole system. The monolithic systems privilege the effectiveness: the implementation lot remains in the addressing space of the core without requiring context switch.

In *microkernels* like QNX [HIL 92], the services are separate processes communicating through IPC (*interprocess communications*). The protection fields acting like units of containment and the reduced size of the kernel make these systems more certain. They are also more flexible: their modular structure enables the dynamic loading of new services. However, the performances are the weak point of these systems because of the cost of the IPC[3].

The *extensible kernels* with single addressing space like SPIN [BER 95] are currently privileged in the embedded systems. Eliminating from the architecture certain abstractions – as the processes to the profit of the *threads* – leads to less expensive context switches. These minimal kernels contain only essential services: the TCB is thus easier to certify. Easily reconfigurable by dynamic code loading in the kernel, these systems are vulnerable to the attacks without additional mechanism of protection[4].

Component-based kernels such as Think [FAS 02] authorize a larger flexibility due to a model of a more homogenous architecture: the integrity of the kernel is made of an assembly of basic units of reconfiguration, the *components*, which are reusable and composable in order to conceive dedicated and minimal systems. The performances obtained are appreciably comparable with those of traditional systems.

The table below summarizes the preceding discussion. We will note that the more the kernel is structured, the more its architecture offers implementation tips for flexibility.

3 The additional cost induced by the mechanism of authorization is weaker when it is located out of the kernel: the addressing space changes are fewer. However, in this case, it is the safety which is weakened.

4 In addition, these systems are often written in programming languages which enable evidence of safety of execution, verifiable before the code downloading in the kernel.

Feature	Monolithic kernel	Microkernel	Extensible kernel	Component-based kernel
Security	– –	+ +/– 4	–/+ + 5	+ +
Performances	+ +	– –/+ 4	+	+
Flexibility	– –	+	+ +	+ + +
Simplicity	– –	+	+ +	+ + +

3.4.2.2. *Security model*

The policies of access explored many security properties starting from confidentiality [BEL 76] or integrity [BIB 77, BOE 85, CLA 87] to the separation of privileges [BRE 89, GLI 02, SIM 97]. The traditional model of authorization known as *access matrix* (AM) [LAM 73] is revealed to be unsuited to prove the safety of the system, the problem being undecidable [HAR 76]. The recent research directions turned toward less expressive models that are also easier to implement [BAD 95]. These models are also less flexible, the rights of access being static. Other works [GLI 02] extend the AM model[5], for example, with constraints facilitating the verification of safety properties. The privileges are managed in a dynamic way. The complexity of expression of these constraints unfortunately makes these models difficult to implement. A compromise is thus necessary between the expressivity of the model measuring its generality and its flexibility, its implementation character and the facility of verifying the properties of safety.

This multiplicity of models reflects an absence of consensus concerning the architecture of a safe kernel, each system supporting its own safety policies. In order to remedy this, neutral mechanisms of authorization can be developed with respect to a model of safety known as *policy-neutral* (PN). No safety policy being optimal, the interest lies in the capacity to support multiple policies and to federate[6] them using the same mechanism. Let us mention the languages [JAJ 97, WIJ 01], the mechanism of Unix process containment [BAD 95] or the support of security models in the form of Linux modules [WRI 02]. In the microkernel [MIN 95], the mechanism of intercepting the access requests is implemented by a PN security

5 Thus, by extending the access matrix with the concept of *role*, the RBAC model [SAN 96] is simpler to manage and can represent several classes of safety policies.

6 For example, in order to control the heterogenity of the networks where the mobile terminal evolves, each network having its own safety policy.

server, whereas the decision function is a service system dependent on a particular model of safety. Other works relate to extensible kernels [GRI 02] or component-based kernels [JAR 04].

3.4.2.3. *Localization of the protection mechanism*

The protection mechanism can be located at the hardware level. Such a solution constitutes a minimal mechanism, impossible to bypass, and strongly protected against failures and attacks. Thus, the memory management unit (MMU) ensures the containment of applications by defining addressing spaces. Let us also mention the mechanisms of exclusion at the level of the hardware bus as in [HAL 03].

Protection can also be carried out using safe languages like Java. These languages do not have the pointer concept and ensure the complete mediation due to the concept of *interface*, which is the only means of access to a material resource. These relatively flexible solutions make it possible to easily carry out access control to fine grain. However, their safety is relatively weak, an attack often leading to a complete takeover of the *run-time* language.

Between these two extremes there is a vast choice of authorization mechanisms of the kernel that can be implemented at the applicative level. Their detailed comparison exceeds the framework of this chapter. In addition to the traditional solutions like the capability systems, we can mention various forms of interposition:

– interception of calls system [BER 02, PRO 03] whose interface is a favored place of the kernel in order to easily ensure the complete mediation;

– authorization tips [WRI 02];

– or container applications [FRA 99].

Other approaches like [KIR 02] provide protection from code injection attacks.

The more the mechanism of protection approaches the kernel, the less the performances will be degraded, the number of context switches between kernel and space user being more reduced. The safety of the system will be increased, the mechanism being less likely to be affected. On the other hand, a reconfiguration will be in general more complex to implement, the kernel having to be modified or recompiled.

3.4.2.4. *Degree of assurance*

The question here is to be convinced that the kernel is worthy of confidence. In addition to the use of test techniques, certain kernels can be the subject of security evaluations [NIS 99, USD 85] going up to a formal validation. The verification techniques used in these evaluations can also be used in order to ensure the safety of the code downloaded in the kernel [NEC 97]. Due to the current limits of the proof techniques, a complex mechanism of authorization will make the system more difficult to certify.

3.4.2.5. *Perspectives*

Among the preceding tracks, the component-based approach seems the best one to succeed in maintaining the delicate balance between reconfigurability and safety. This simple architecture model authorizes the creation of various types of kernels without really penalizing performances. Its modular character makes it possible to explore several safety models and to place the protection mechanisms (access control, detection of intrusion) at various levels of abstraction. Lastly, its capacity to create minimal kernels should make it possible to reach high levels of assurance, which gives confidence. This architecture makes it possible to simply adapt the system to the changes that are likely to intervene during its life cycle, without truly endangering its safety, the component being a unit of safety and reconfiguration at the same time. This approach should thus develop in the years to come for the design of reconfigurable terminals.

3.4.3. *Reconfigurability and performances*

3.4.3.1. *Introduction*

A reconfigurable system operates under two modes:

– a "nominal" mode when no reconfiguration is taking place;

– a "reconfiguration" mode.

The operations involved in the reconfiguration mode were previously detailed (section 3.2.3.2) but at present we do not have tangible background information on the cost induced by the latter, in terms of resources. Let us quote as an example the additional costs related to the following operations:

– the cost of mechanisms making it possible to detect a stable state in order to engage reconfiguration (the existing mechanisms are founded either on a thread model, or on the systematic calculation of the incoming and outgoing calls in each component: a first evaluation doubles the call duration method for the components of the first hierarchical level);

– the cost in resources (primarily memory) monopolized by the resident code intended to control the reconfiguration (code for the instantiation activation of the new components and the edition of new connections);

– the complementary operations like the recovery of resources used by an eliminated component (undoubtedly at a considerable cost, for example, with respect to memory management);

– costs for the reconfiguration operation itself (instantiation of components, revocations/link editings, etc.).

We will focus on the nominal mode. The main additional cost related to the reconfigurable character of a system comes from the two following constraints:

– the need to store the necessary structure information (capacities of introspection of a component) and to manage their access;

– the capacity of modifying the link chains.

3.4.3.2. *Reconfiguration impact on nominal mode performances for a hierarchical "component-based" system*

We will detail here some evaluation elements inspired by an experiment currently developing in our laboratories: the project of the component-based system THINK [FAS 02]. The impact of a "component-based" design in terms of memory consumption and calculation speed is evaluated below, with respect to a traditionally built system (for example, by C compilation and traditional link editing). The following evaluation relates to the control autoconsumption in terms of resource memory. A distinction is made between components of the lowest hierarchical level (primitive components) and the others (composite component; see section 3.2.3.2).

Primitive components		Composite components	
Interfaces	Control autoconsumption, expressed in bytes and by number of interfaces or attributes (n)	Entities	Control autoconsumption, expressed in bytes and by number of entities (n)
Server	$4 + 12 \times n$	Subcomponent	$4 + 8 \times n$
Client	$4 + 8 \times n$	Internal link	$4 + 12 \times n$
Attributes	$4 + 8 \times n$	Attribute assignment	$4 + 12 \times n$

Examples:

– an IP stack of 50 ko (primitive component) presents a 2% increase in print memory;

– for a 1 ko visualization buffer (primitive component) the increase is 1%;

– for a TCP/IP stack (composite component), consisting of 10 subcomponents interconnected by 25 links, the excess of conceded size is up to 400 bytes[7].

In our experimentation, the impact on the calculation speed comes from the implementation of interfaces which, on ARM architecture, "costs" 5 additional instructions, which is comparable to a method call (from 2 to 3 instructions). Since the method call has to be preceded by an access to the related interface, the magnitude order in the worst case would be lower to the double.

The "memory and speed" costs proposed here are costs at the level of the component or, more exactly, of the method call and thus at a very low level. The estimate of the total costs at the level of a complete system cannot be carried out by simple extrapolation of these elementary costs but must imperatively take into account the size of the components in order to obtain a relevant estimate: the "large" component-based systems will be less penalized than the "small" component-based systems, the latter offering more possibilities of reconfiguration.

7 The figures proposed here should be considered with caution: their significance is limited by the existence of many optimization possibilities not yet explored, or simply owing to the fact that they concern a single experiment.

In conclusion, the impression left from these first elements resulting from our experiment of component-based design is therefore relatively reassuring with regard to the criterion of speed: the reconfigurable systems do not seem much slower than their non-reconfigurable version. The memory footprint criterion will have to be refined in order to take into account the control mechanisms of reconfiguration but the function code, which is only used in nominal mode, does not seem very demanding with respect to the function code of a non-reconfigurable version.

3.5. Some open problems

3.5.1. *The problem of reversibility*

A particularly significant case of software reconfiguration refers to the adaptation of functionalities in order to answer to the evolution of the terminal's external conditions (network size, deterioration of the transmission quality), or the evolution of the availability of the internal resources (battery level). The goal required in this case is an optimization of the instantaneous performances. When, after reconfiguration, the optimization sought is not or not longer reached, it is essential to be able to return to a nominal state (typically, the state preceding the reconfiguration). If necessary, this operation (just as the reconfiguration operation) must be carried out without service interruption.

This scenario opens the problem of reversibility of reconfigurations and brings again into question the compromise between performances and reconfigurability: the search for performance (typically, by replacing the indirections by "cabled" references) can result in eliminating further possibilities of reversibility by solidifying certain links between components.

However, beyond this consideration, the implementation of returning to a former configuration is far from being trivial. We can imagine that during the life of a reconfigurable system, it will undergo a certain number of successive reconfigurations. An opposite operation of "retroconfiguration" corresponds to any reconfiguration operation, which brings back the system to the origin configuration. One of the possible solutions is to apply a series of successive reconfigurations to the system, "opposite" to the reconfigurations carried out. This operation requires to store all the successive modifications separately. During the complex reconfiguration sequences, the strategy of the "opposite sequence" is not realistic, for obvious reasons of volume of data to be stored, or duration. Consequently, the problem of research and implementation of an optimal sequence of reconfigurations in order to return to the desired configuration arises. It is about a particular case of a more general problem consisting of finding an optimal sequence of reconfigurations (in terms of resources) in order to pass from an architecture to another.

3.5.2. *The problem of continuity of service*

The problem of continuity of service arises as soon as it is a question of adaptation to a context switch (radio context, availability of the terminal resources, etc.). In practice, it is about a difficult problem, which is intrinsically related to the execution reconfiguration. In this section, we will give some tips that help in foreseeing the problem.

We place ourselves in the (simple) case of replacing a component by another component. A protocol of reconfiguration with continuity of service is characterized by the two following dimensions:

– latency (time period between the decision and the end of reconfiguration operation). The need to suspend the activity of the component to be reconfigured is reflected gradually on the set of components directly or indirectly related to the latter. Taking into account the latency implies temporary data storage and the resynchronization of the processing at the end of the reconfiguration;

– disturbance of quality of service (loss of data, desynchronization in data processing, loss of history, etc.).

We have seen that a reconfiguration operation required the implementation of several operations independent to a certain extent from one another: instantiation of the replacing component, state transfer between the two components, connection of the replacing component, activation of the replacing component, disconnection of the old component, recovery of the resources of the old component, etc. The relative independence between these operations makes it possible to imagine several very different scenarios, like for example:

– scenario 1:

(i) closedown/disconnection of the former component,

(ii) recovery of the resources of the former component,

(iii) instantiation/connection/activation of the new component;

– scenario 2:

(i) instantiation/connection/activation of the new component,

(ii) operation of the two components in "parallel",

(iii) rocking between the two components,

(iv) closedown/disconnection of the former component,

(v) recovery of the resources of the former component.

The first scenario has the advantage of reducing latency, whereas the second limits the disturbances by favoring a more fluid transfer of state: the parallelization (in background task) of the new component during a sufficient time period enables the initialization of its state, even before the rocking takes place between the two components. However, the various operations are constrained by the capacities of the host infrastructure. For example, for reasons of availability of resources, it can be impossible to activate in parallel the old and the new component (see scenario 2).

In short, the various combinations between elementary operations lead them to different compromises in terms of latency, service disturbance, but also of costs in resources. The problem of adjusting this compromise is still very largely open today and can be regarded as one of the "major problems of infrastructure" concerning the reconfigurable radio.

3.6. Conclusion

The reconfigurable radio and its terminal application is a very active field today and many paths are followed by researchers worldwide. At the crossroad of multiple elementary disciplines, several approaches of the reconfigurability of the terminal were followed in this chapter in order to show that there is not only one way leading to the reconfigurable terminal, but a multitude of paths, all being the result of trade-off at various levels.

3.7. Bibliography

[BAD 95] BADGER L., STERNE D., SHERMAN D., WALKER K., HAGHINGHAT S., "Practical Domain and Type Enforcement for Unix", *IEEE Symposium on Security and Privacy (S&P'95)*, 1995.

[BEL 76] BELL D., LAPADULA L., Secure Computer System: Unified Exposition and Multics Interpretation, Technical Report ESD-TR-75-306, Electronics Systems Division, Bedford USAF Base, Department of Defense, 1976.

[BER 95] BERSHAD B., SAVAGE S., PARDYAK P., SIRER E., FIUCZYNSKI M., BECKER D., CHAMBERS C., EGGERS S., "Extensibility, Safety and Performance in the SPIN Operating System", *ACM Symposium on Operating System Principles (SOSP'95)*, 1995.

[BER 02] BERNASCHI M., GABRIELLI E., MANCINI L., "REMUS: A Security-Enhanced Operating System", *ACM Transactions on Information and System Security (TISSEC)*, vol. 5, no. 1, p. 36-61, 2002.

[BIB 77] BIBA K., Integrity Considerations for Secure Computer Systems, Technical Report ESD-TR-76-372, Electronics Systems Division, Bedford USAF Base, Department of Defense, 1977.

[BOE 85] BOEBERT W., YAIN R., "A Practical Alternative to Hierarchical Integrity Policies", *National Computer Security Conference*, 1985.

[BRE 89] BREWER D., NASH M., "The Chinese Wall Security Policy", *IEEE Symposium on Security and Privacy (S&P'89)*, 1989.

[BRU] BRUNETON E., COUPAYE T., STEFANI J.B., "The Fractal Component Model", http://fractal.objectweb.org.

[CIB 04] CIBAUD D., "Functional study of FDD equalizer in the context of TDD reconfiguration", PIMRC 2004, Barcelona, 2004.

[CLA 87] CLARK D., WILSON D., "A Comparison of Commercial and Military Computer Security Policies", *IEEE Symposium on Security and Privacy (S&P'87)*, 1987.

[CZY 02] SZYPERSKI C., *Component Software*, 2nd edition, Addison-Wesley, 2002.

[DEN 02] DENYS G., PIESSENS F., MATTHIJS F., "A Survey of Customizability in Operating Systems Research", *ACM Computing Surveys*, vol. 34, no. 4, p. 450-468, 2002.

[EKR 94] EKROOT L., DOLINAR S., "Maximum-Likelihood Soft-Decision Decoding of Block Codes Using the A* Algorithm", *The Telecommunications and Data Acquisition Progress Report 42-117, Jet Propulsion Laboratory*, p. 129-144, 15 May, 1994.

[ESM] http://www.esmertec.com.

[FAS 02] FASSINO J.P., STEFANI J.B., LAWALL J., MULLER G., "Think: a Software Framework for Component-Based Operating System Kernels", *USENIX Annual Technical Conference (USENIX'02)*, 2002.

[FRA 99] FRASER T., BADGER L., FELDMAN M., "Hardening COTS Software with Generic Software Wrappers", *IEEE Symposium on Security and Privacy (S&P'99)*, 1999.

[FEL 02] FELDMAN J., FRIGO M., ABOU-FAYCAL I., "A Fast Maximum-Likelihood Decoder for Convolutional Codes", *IEEE Vehicular Technology Conference (VTC)*, fall 2002.

[GCJ] http://gcc.gnu.org/java.

[GLI 02] GLIGOR V., GAVRILA S., FERRAIOLO D., "On the Formal Definition of Separation of Duty Policies and their Composition", *IEEE Symposium on Security and Privacy (S&P'98)*, 1998.

[GRI 02] GRIMM R., BERSHAD B., Providing Policy-Neutral and Transparent Access Control in Extensible Systems, Technical Report UW-CSE-98-02-02, 2002.

[HAL 03] HALFHILL T., "ARM Dons Armor: TrustZone Security Extensions Strengthen ARMv6 Architecture", *Microprocessor Report*, document available at http://www.arm.com/miscPDFs/4136.pdf, 25 August 2003.

[HAR 76] HARRISON M., RUZZO W., ULLMAN J., "Protection in Operating Systems", *Communications of the ACM*, vol. 19, no. 8, 1976.

[HAN 02] HAN Y.S., CHEN P.N., WU H.B., "A Maximum-Likelihood Soft-Decision Sequential Decoding Algorithm for Binary Convolutional Codes", *IEEE Trans. on Communications*, p. 173-178, February 2002.

[HEI 01] HEINEMAN G.T., COUNCIL W.T. (ed.), "Component-Based Software Engineering", *Putting the Pieces Together*, Addison-Wesley, 2001.

[HIL 92] HILDEBRAND D., "An Architectural Overview of QNX", *USENIX Workshop on Micro-Kernels and other Kernel Architectures*, 1992.

[JAJ 97] JAJODIA S., SAMARATI P., SUBRAHMANIAN V., "A Logical Language for Expressing Authorizations", *IEEE Symposium on Security and Privacy (S&P'97)*, 1997.

[JAR 04] JARBOUI T., FASSINO J.P., LACOSTE M., "Reconfigurable Access Control for Component-Based OS Kernels", *E2R Workshop on Reconfigurable Mobile Systems and Networks beyond 3G, IEEE International Symposium on Personal, Indoor and Mobile Radio Communications (PIMRC'04)*, 2004.

[JAV] http://www.sun.com/software/java.

[KIC 91] KICZALES G., RIVIÈRES J., BOBROW D.G., *The Art of the Metaobject Protocol*, Chapters 5 and 6, Cambridge MIT Press, 1991.

[KIR 02] KIRIANSKY V., BRUENING D., AMARASINGHE S., "Secure Execution *via* Program Shepherding", *USENIX Security Symposium*, 2002.

[LAM 73] LAMPSON B., "A Note on the Confinement Problem", *Communications of the ACM*, vol. 16, no. 10, p. 613-615, 1973.

[LEA 00] LEAVENS G., SITARAMAN M. (ed.), *Foundations of Component-Based Systems*, Cambridge University Press, 2000.

[LOS 99] LOSCOCCO P., SMALLEY S., "Integrating Flexible Support for Security Policies into the Linux Operating System", *USENIX Annual Technical Conference (USENIX'99)*, 1999.

[MIN 95] MINEAR S., "Providing Policy Control over Object Operations in a Mach-Based System", *USENIX Security Symposium*, 1995.

[MUH 04] MUHAMMAD K., "A discrete-time Bluetooth receiver in a 0.13 µm Digital CMOS process", *IEEE International Solid-State Circuits Conference*, 2004.

[NEC 97] NECULA G., "Proof-Carrying Code", *ACM Symposium on Principles of Programming Languages (POPL'97)*, 1997.

[NIS 99] NATIONAL INSTITUTE OF STANDARDS AND TECHNOLOGY, *Common Criteria for Information Technology Security Evaluation (CC)*, 1999.

[OCA] http://www.caml.inria.fr/ocaml.

[PAT 87] MAES P., "Concepts and experiments in computation reflection", *ACM SIGPLAN Notices*, 22(12), p. 147-155, December 1987.

[PRO 03] PROVOS N., "Improving Host Security with System Call Policies", *USENIX Security Symposium*, 2003.

[SAN 96] SANDHU R., COYNE E., FEINSTEIN H., YOUMAN C., "Role-Based Access Control Models", *IEEE Computer*, vol. 29, no. 2, p. 38-47, 1996.

[SIL 98] SILBERSCHATZ A., GALVIN P., *Operating System Concepts*, Addison-Wesley, 1998.

[SIM 97] SIMON R., ZURKO M., "Separation of Duty in Role-Based Environments", *IEEE Computer Security Foundations Workshop (CSFW'97)*, 1997.

[USD 85] US DEPARTMENT OF DEFENSE, *Trusted Computer Security Criteria (TCSEC)*, 1985.

[WAL 99] WALDEN R.H., "Analog-to-Digital Converter Survey and Analysis", *IEEE Journal on Selected Areas in Communication,* April 1999.

[WIJ 01] WIJESEKERA D., JAJODIA S., "Policy Algebras for Access Control – The Propositional Case", *ACM Conference on Computer and Communications Security (CCS'01)*, 2001.

[WRI 02] WRIGHT C., COWAN C., SMALLEY S., MORRIS J., KROAH-HARTMAN G., "Linux Security Modules: General Security Support for the Linux Kernel", *USENIX Security Symposium*, 2002.

Chapter 4

A UMTS-TDD Software Radio Platform

4.1. Introduction

This chapter describes the concrete design of a reconfigurable mobile radio system, based on a UMTS-TDD software radio platform.

Why develop such reconfigurable real-time radio platforms? On the one hand, the use of software radio technology makes it possible to keep the costs of the equipments reasonable because they are mainly based on inexpensive equipments such as PC microcomputers. On the other hand, this provides the ideal means to carry out significantly large mobile radio system experiments. Finally, these platforms are used in order to validate theoretical progress, be it in signal processing, in resource allocation strategy or in the network field, especially in the very promising area of the beyond-3G (B3G) wireless networks. As an example, we can mention:

– the management of radio resources in heterogenous wireless networks containing several types of access technologies (WLAN, UMTS), based on the IPv6 protocol;

– the smart-antenna technologies;

Chapter written by Christian BONNET, Hervé CALLEWAERT, Lionel GAUTHIER, Raymond KNOPP, Pascal MAYANI, Aawatif MENOUNI HAYAR, Dominique NUSSBAUM and Michelle WETTERWALD.

– the allocation and scheduling of radio resources;

– the management of quality of service (QoS).

The developments presented in this chapter and also made available for the researcher community on the [EUR] website, are based on the convergence of three technological trends:

1. Collaborative radio standards: 3GPP [3GP], IEEE 802.x [IEE];

2. New generation Internet technology for the mobiles – IPv6 [IP], IP mobile [MIP], SIP [SIP];

3. Open source software and operating systems – Linux [LIN], RTLinux [RTL], RTAI [RTA].

The availability of software that complies with the version 4 of the UMTS-TDD standard (at least for the standard radio access part), under a GNU *general public license* (GPL), makes it possible to promote the academic collaboration in the area of B3G open systems. This software can be used stand-alone (non-real-time) or with RF radio frequency equipment (real-time), which is also available from the [EUR] website. We therefore hope that this opening policy will lead to new collaborative projects based on our platforms, present and future.

Those are or have been used in several projects. The French projects are funded by the French Ministry of Research under the RNRT label [RNR], i.e. SAMU, PLATON, ERMITAGES, @IRS++, RHODOS, LAO-TSEU, COSINUS. At the European level, an UMTS-TDD/IPv6 platform was part of the IST Moby Dick [MOB] project (fifth Framework Program), which proposed innovative methods for the direct interconnection of a 3GPP radio interface with an IPv6 core network. It is now used in projects of the next Framework Program following (FP6), among which we find WIDENS, DAIDALOS and E2R [IST].

As a consequence of these collaborative projects, several experimental B3G networks have been or are being rolled out, notably in Germany (University of Stuttgart) and in France, in Sophia Antipolis (Eurecom institute) and in Paris (Motorola Research Center). In these various sites, UMTS-TDD frequency bands have been assigned to the research activities associated with these projects.

This chapter describes the various components of the software radio platform and their functionalities. The following section discusses the various material components (radio frequency and acquisition) designed for radio gateways (base stations) and for the reconfigurable user terminals. We will then describe the software technologies implemented in these platforms (real-time operating systems, signal processing programming, simulation environment and free software design of the radio protocols). Finally, we will explain how the innovations in the wireless network protocols enabled the experimentation of a pure-IPv6 heterogenous infrastructure, as well as reconfigurability capabilities of the platform.

4.2. Hardware architecture

4.2.1. *Radio gateways*

The current radio equipment is designed to operate in a *time division duplex* (TDD) mode in a 5 MHz bandwidth channel. It operates in all the TDD bands of the IMT-2000 spectrum, between 1,900 and 1,920 MHz, but other frequency bands can possibly be supported if necessary.

The base stations, called *radio gateways* (RG), can be configured in one of the two following power classes:

– low power picoBTS (25 dBm/antenna element, – 40 dBc ACLR, 7 dB noise figure);

– high power microBTS (34 dBm/antenna element, – 45 dBc ACLR, 5 dB noise figure).

These two configurations include one radio frequency (RF) module and one PCI acquisition card per antenna element. Moreover, the high power configuration requires a front-end power amplifier and a low noise amplifier. They can be equipped with one or several antennae in order to benefit from space diversity.

The basic RF module for the RG has been designed together with Philips Semiconductors in Sophia Antipolis and is in the form of a 19 inch rack (similar to an Ethernet switch). The transmission functions are comparable to those of a high-end digital signal generator (14 bit sampling). In addition, it brings the capability of a real-time operation when it is controlled by a RTLinux PC. A digitally tunable local oscillator runs an up-conversion towards the 1.9 GHz frequency. The digital control also makes it possible to obtain several gain levels. The reception capabilities are noticeably comparable to those of a good signal analyzer (14 bit sampling), also here with the added capability of a real-time operation. Sampling clocks can be synthesized onboard or obtained from an outer source. This module is represented with its input/output functions in Figure 4.1.

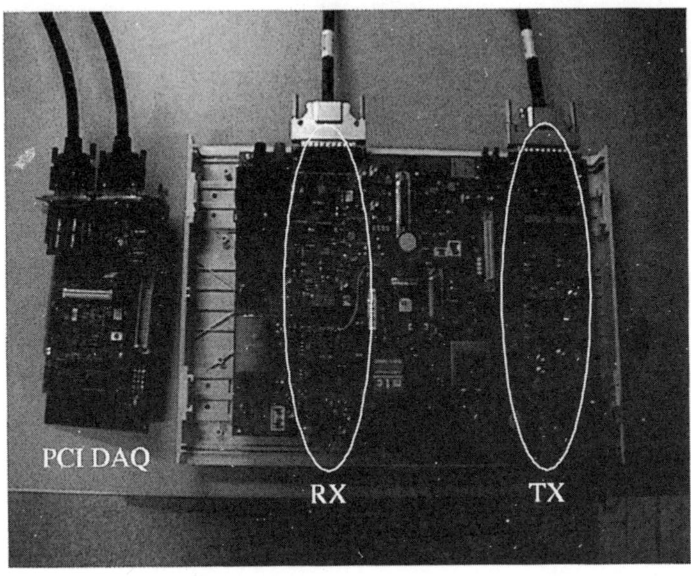

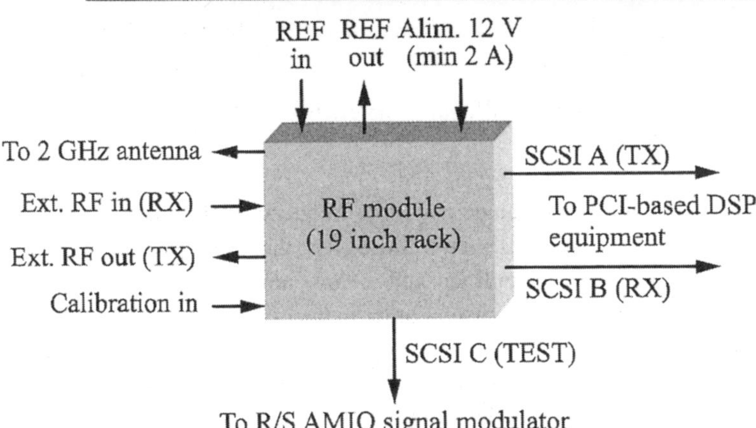

Figure 4.1. *Basic RF module*

The module has been designed in view of obtaining maximum flexibility and reconfigurability. It enables a stand-alone operation either with a 2 GHz antenna or with a high power/low noise external module connected by low-loss cables at the external RF input/output (I/O) ports. These configurations are shown in Figure 4.2.

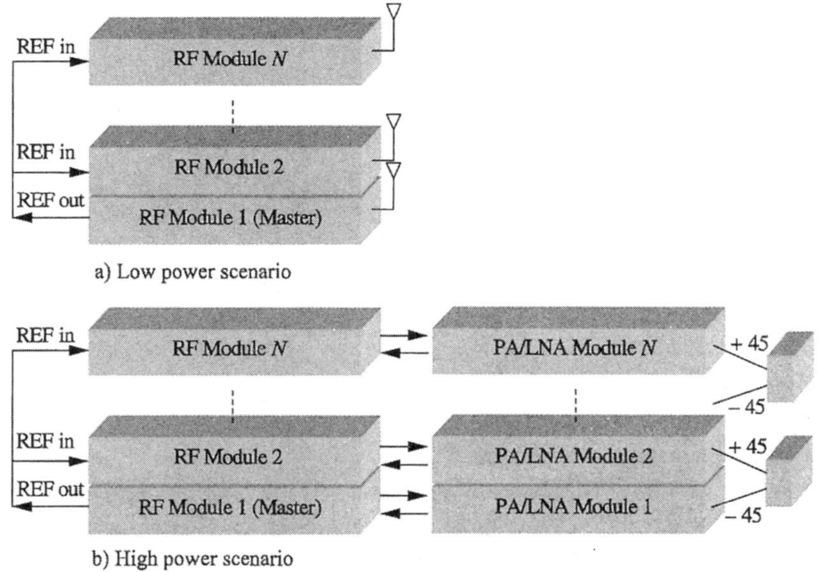

Figure 4.2. *Low or high power multiantenna configurations*

This module also enables a multiple-antenna operation and synchronized clocks, by means of the same RF I/O ports. Power supply is provided by a standard 12V power generator (minimum 2 A). The RF module can therefore be plugged on a car battery in order to test the terminal in a high mobility environment.

The pictures in Figure 4.3 illustrate two examples of radio gateway deployment. The picture on the left shows a picoBTS designed from a PC and a RF module connected to only one antenna. This configuration is especially adapted to a demonstration in an indoor environment, where the levels of transmitted power are quite low. The picture on the right illustrates a configuration in an outdoor environment. We see an antenna installed on the roof of the building and connected to a high power/low noise amplification module (PA/LNA), which is circled at the bottom of the picture. The radio gateway and the RF module are not visible because they are located in the office under the antenna.

The acquisition/generation of the digital signal is provided by a specific card connected on the PCI (*peripheral component interconnect*) bus of a microcomputer. It contains a FPGA (*field programmable gate-array*) as well as a traditional PCI bus controller. Some ultra-SCSI standard cables provide the connectivity with the RF modules. The FPGA is only used for the acquisition of the signal and all the signal processing is then executed by software in the PC.

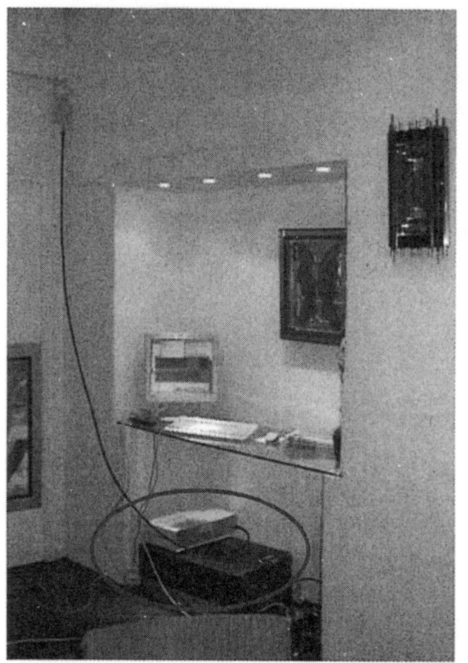

RF module + PC

PA/LNA module
(installation on the roof)

Figure 4.3. *Examples of radio gateway rollout*

4.2.2. *Terminal equipment*

The same RF equipment can also be used for the user terminals after a simple modification in the associated software. A new generation of RF modules, which are adapted to mobile terminals, is under development based on a state-of-the-art FPGA and enabling both acquisition/generation and the front processing of the signal (filtering, DFT, resampling). These modules also provide a CardBus interface for laptops, the panel-PCs or the embedded PCs. The RF part includes both a 1.9 GHz transceiver and a wideband tunable front-end (3-6 GHz). It can operate with one or two antennae as one chooses.

This equipment is designed as part of the research projects on *ad hoc* networks targeted for the communications of the public safety organizations (firemen, etc.). It will thus provide both the spectral agility and the reconfigurability while remaining in a small form-factor. A prototype is shown in Figure 4.4.

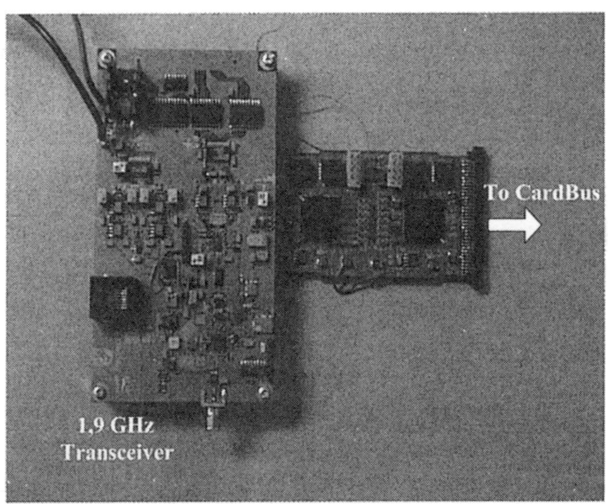

Figure 4.4. *Prototype for a reconfigurable terminal*

4.2.3. *RF emulation*

With the purpose of developing and testing the radio and network access protocols in real-time, without necessarily deploying a real transmission in carrier frequency, it is possible to use a connected RF emulation environment operating in intermediate frequency (70 MHz). This emulator brings all the real-time functionalities while avoiding the potential problems related to the deployment of a live network (propagation and interference problems, etc.) and to the RF subsystem (licensing, etc.). This equipment gathers in a same structure four analog/digital conversion cards that correspond to four RF modules without the radio part. Each card can be connected to a PCI acquisition card in the same way as in the standard RF system, which corresponds to the input point in the PC microcomputer. Several configurations can therefore be created (for example, 2 RG and 2 terminals or 1 RG and 3 terminals, etc.).

4.3. Software architecture

4.3.1. *RTLinux*

The real-time software runs in an RTLinux/GPL operating system designed by FSMLab [RTL]. It enables a real-time execution. Since Linux contains many innovative network functionalities such as IPv6, mobile IPv6, routing and video-transmissions it provides an ideal platform for developing experimental B3G network equipments.

RTLinux [RTL] is an extension of the Linux operating system that supports real-time interrupt handlers and periodic tasks, with interrupt latencies and a scheduling jitter that are close to the hardware limits. In the worst case, the interrupt latency on a standard PC is lower than 15 microseconds from the time the hardware interrupt has been asserted.

How does RTLinux operate? The fundamental idea is to replace the code that activates/deactivates usual interrupts by a code that activates/deactivates soft interrupt. These hard interrupts are then all handled by the RTLinux real-time micro-kernel. Due to this separation, the Linux kernel cannot disable the interruptions of the real-time threads which remain under the control of RTLinux. The Linux device drivers then work as usual, provided that they have been correctly written and they do influence the real-time threads handled by RTLinux. We can then include purely real-time tasks in the system while guaranteeing their performances and this without impacting the normal Linux tasks (excluding execution speed).

Although it is not strictly POSIX-compliant (*portable operating system interface*: the Unix standard of the IEEE that specifies the core of the system), RTLinux provides a similar interface, i.e.:

– a real-time thread library (*pthreads*);

– schedulers;

– the handling of external interruptions;

– SMP (*symmetric multi-processing*) functions.

The protocol radio layers described below as well as the device drivers specific to the RF equipment for the platform have been developed in this environment.

4.3.2. *Programming modes of the processor*

The software is written essentially for the C GNU GCC compiler, but some signal processing algorithms directly use the Intel SIMD instructions, fixed point and floating point (MMX, SSE, SSE2) in the purpose of obtaining the maximum efficiency with the Intel Pentium processors used. Other encoding functions have been written by using the free Intel C compiler for Pentium 4 [INT]. The software also includes signal processing functions optimized for the UMTS modems, such as matched filtering, pulse-shape filtering, an FFT (fast Fourier transform) based channel estimation and Viterbi and turbo channel decoders.

4.3.3. *RF simulation software*

Discrete-time propagation channels have been modeled, thus making it possible to use in simulation the same source code as the real-time operation, which gives the possibility of rapidly testing new algorithms on the various protocols.

This simulation mode is designed in order to operate with a network of PC microcomputers, although a single machine is sufficient for the minimal configuration. The ideal operation is nevertheless a cluster of Linux PC computers. The compilation, either for the simulation environment or real-time execution, is controlled by a simple directive of the Makefile. In the case of the simulator, the code that is specific to the RF equipment is not used. The software is then run in the user space rather than in the kernel space, while maintaining the same multi-threaded structure as the real-time code. Signals are exchanged between a set of RGs (potentially on different machines) and a set of terminals (also on different machines). The exchange is executed by means of a TCP/IP client/server socket interface and a third entity that controls the signal flow, which we call a channel server. This architecture is shown in Figure 4.5. The channel server is a multi-threaded process, typically running on a powerful machine. It simulates the RF part of the radio network and transfers the digital signals among the entities of this network. The TCP/IP sockets directly interface the low layers of the radio protocol stack, those which communicate with the hardware subsystem during the real-time execution. Each thread of the channel server controls the signals emanating or arriving at an antenna in the network. Each node of the network, either RG or MT, can contain several antennae (represented by several connections between the nodes and the channel server; see Figure 4.5), in order to simulate a space-time processing functionality. The channel server requires some configuration information: the locations of the nodes in the network, the path-loss and fading model parameters, as well as the transmission power levels to be simulated.

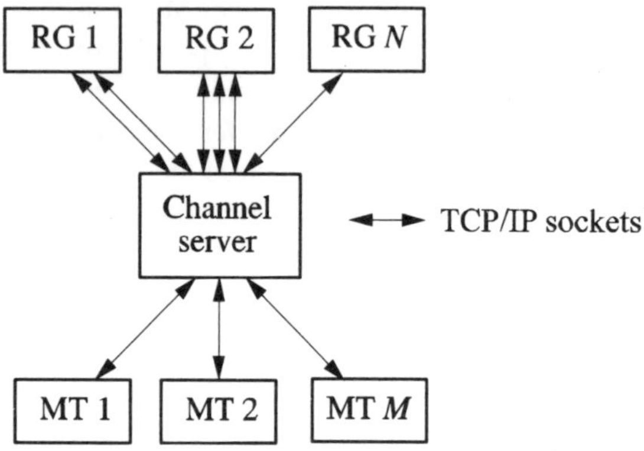

Figure 4.5. *RF simulation environment*

4.4. Connection to the IPv6 network

4.4.1. *"Pure-IPv6" architecture*

Figure 4.6 shows the evolution of UMTS network entities towards an architecture based on an IPv6 core network. We can see in this figure that bringing IP to RG has a significant impact on the overall architecture of the network. In fact, we short-circuit some of the UMTS networking entities (RNC/SGSN/GGSN), which leads to a significant reduction of the number of bearer services. A bearer service enables the provision of a contracted QoS using the services of the lower protocol layers.

The radio bearer service between the mobile terminal (MT) and the RG remains unchanged because it defines the configuration of the *access stratum* for this service. The UMTS bearer service is also *a priori* necessary in order to enable the connection between the UMTS subsystem and the IP world. However, other services are omitted in the proposed architecture because the corresponding interfaces do not exist. The system is therefore very simplified.

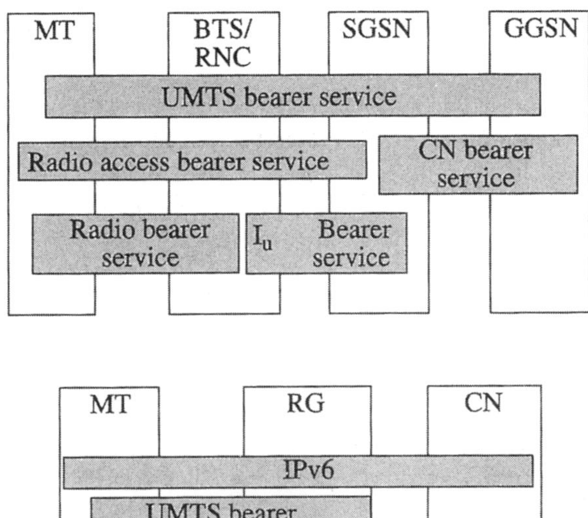

Figure 4.6. *Evolution of UMTS entities towards a pure-IPv6 network*

4.4.2. *Radio protocols*

The direct interconnection between the UMTS-TDD and the IPv6 is carried out by means of the protocol stack represented in Figure 4.7 and adapted from [23.060].

The physical layer is divided into two sublayers: sublayer L1L (*PHY-low*) and sublayer L1H (*PHY-high*). Sublayer L1L interfaces directly with the acquisition card and runs the first level of signal processing (matched filtering, pulse-shape filtering, channel estimation and synchronization). It operates on a slot-by-slot basis (666 μs in the case of 3GPP modems). Sublayer L1H runs the channel encoding/decoding and interleaving. It operates on a frame-by-frame basis (10 ms in the case of 3GPP modems). Layer 2 contains the MAC (*medium-access control*) and RLC (radio link control) control layers, which are responsible for the PDU (*packet data units*) scheduling on the transport channels, for the segmentation of packets and for the retransmission protocols. It is completed by the PDCP (*packet data convergence protocol*) layer, which is the access point for the data traffic toward the RLC layer. It also operates on a frame-by-frame basis. The implementation of layers 1 and 2 complies with the 3GPP specifications, except for some transport channels that are not supported.

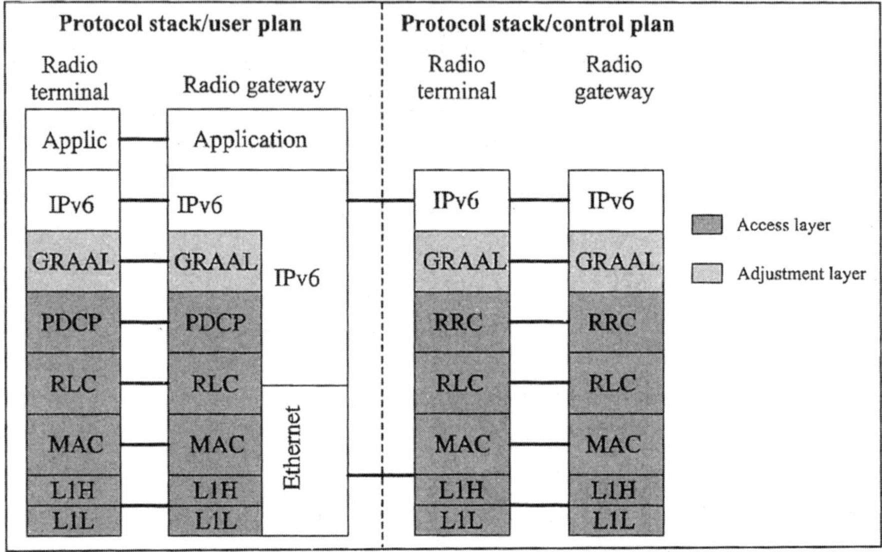

Figure 4.7. *Modified protocol stack*

The level 3 protocol layer of the software radio platform deviates from the 3GPP standards because it provides direct interconnection with an IPv4/IPv6-based network. It includes a control function (RRC – *radio resource control*) that offers an interface to set up and release the radio bearers for the signaling and the data packets, and an abstraction layer (GRAAL – *generic radio access abstraction layer*). This provides the interface and the adaptation between the IPv6 signaling and user traffic mechanisms and the mechanisms that are specific to the radio access of the 3GPP for the support of the mobility functionalities, the call admission, etc.

The radio gateway (RG) is not only a node-B (or base station), which provides the physical layer of the *access stratum* and a part of the MAC layer. The other layers (MAC, RLC and upper ones) are usually located in the RNC (*radio network controller*). The RG as we define it is represented in Figure 4.7. We can therefore consider it as a result of the addition of a node-B and a subset of the RNC. This, among others, enables the real-life deployment of innovative "pure-IPv6" solutions for experimental heterogenous B3G/WLAN wireless networks. The *radio resource management* (RRM), responsible for the management of mobility, the QoS and the configuration of *access stratum* is defined in an Internet type context and no longer in a UMTS one. This architecture enables an interconnection without discontinuity of the various radio accesses all of which operate in direct relation with the IP protocol layer (for example, WLAN and UMTS).

Currently, we have developed the 3.84 MChip/s mode of the UMTS/TDD. The HSPDA versions (high speed packet access) of the version 5 and 1.28 MChip/s (TD-SCDMA) [25.201] are currently under development as part of several public research projects (RHODOS, LAO-TSEU) [RNR]. We also planned to extend the development in order to support new radio and MAC protocols within the frame of *ad hoc* networking of the communications intended for the public safety organization (firemen, etc.) through large and reconfigurable bands. These systems follow the MESA [MES] specifications. Our radio equipment remains compatible with these extensions, which will be created by adding only software functions.

4.4.3. *Interconnection layers*

Two protocol layers contribute to the direct interconnection with IPv6: the GRAAL and the RRC.

The GRAAL is developed as an extension of an IPv4/v6 device driver. When the IP packets enter the GRAAL, they go through a classifier that separates signaling packets and data packets. Signaling packets are either processed locally or sent to the signaling channels, depending on their nature.

For data packets, the GRAAL uses the information of QoS contained in the IP header in order to route the packets to the appropriate radio bearer. A UMTS cell is here considered as an IPv6 subnet and therefore the RG is a router for the users in subnet/cell.

According to the standards, the operator and/or the implementation are responsible for the mapping between the application QoS parameters and the 3GPP bearer service attributes. The GRAAL updates a connection table between the level 2 (UMTS-based) and the level 3 service classes (IP-based).

If there is no existing radio bearer corresponding to a received packet, the GRAAL triggers the establishment of a new channel. When the resources available in the network have been considered as sufficient, it obtains the authorization to establish this new radio channel. It determines the mapping between the DiffServ codepoint contained in the header of the first IP packet and one of the service classes defined for the radio interface. The parameters and resources for the *access stratum* are requested at the RRM and the authentication and the accounting at the responsible entities in the core network. If a radio bearer is already established, the IP packet is directly transferred to the corresponding PDCP/RLC entities.

The GRAAL also provides the following functions in the "control" plan:

– the connection establishment/release procedure of the mobile terminal;

– the establishment/release procedure of the radio bearer;

– the broadcasting (in the radio gateway) of messages like "*neighbor discovery*" (search for neighboring mobiles) and router advertisement;

– the list of adjacent cells;

– the list of signal quality of the adjacent cells (in the mobile stations for the IPv6 *handover*);

– the computation of the IPv6 link-local address.

The RRC layer developed in the platform is a subset of the protocol defined in the 3GPP standard [25.331]. It provides the interfaces and the signaling procedures between the low-layers of the access protocols (PHY/MAC/RLC), the networking protocols (broadcast, paging, connection management) and the RRM-like specific entities. It specifically provides procedures for:

– the establishment and release of the connection to the network;

– the establishment and release of the radio bearers;

– the mechanisms for the paging and broadcast of the system information;

– the collection and the report of the measurements (for example, power levels, signal quality and volume of traffic transferred).

4.4.4. *Management of the radio resources*

A function that is usually not present in the base stations (but is present in the RNC) is the RRM whose algorithms must manage the radio resources of several adjacent cells. This entity is present in the platform.

The RRM is centrally implemented across a group of RGs, which enables a joint management of the radio resources for different cells and, potentially, across different access technologies. In the pure-IP environment considered here, the RRM entity is located in a server of the core network that provides (or will provide in the short-term future) the following functions to the RG clients that it manages:

– automatic configuration of the access stratum based on QoS-based service requests;

– interference mitigation (power control, channel dynamical allocation);

– low level mobility management (for example, soft handover control);

– joint resource allocation across several access technologies (UMTS/WLAN);

– collection and measurement processing algorithms.

4.5. Reconfigurability

4.5.1. *Functional reconfigurability*

The platform is reconfigurable according to several points of view. Two of them have already been addressed. The first one is the reconfigurability from an operational point of view. Since the set of protocol layers is software implemented, it is possible to add, to modify or to suppress some functionalities on the platform depending on the needs of the collaborative project that used it. For example, functionalities associated with the mobility and the broadcasting of multimedia content will be added to the base platform for the IST Daidalos [IST] project. Optimization algorithms of the radio resources within the frame of end-to-end service warranty will be added on the same basic platform for the RNRT Rhodos [RNR] project. New algorithms optimizing the MAC layer for the support of public safety networks are going to be implemented within the framework of the WIDENS [IST] project. For all these projects, the basic software source code is modified and improved by means of new software modules. The hardware part which is based on the RF modules described above remains unchanged.

4.5.2. *Operational reconfigurability*

The second possibility of reconfiguration of the platform concerns the radio part and the packet transport layers (MAC, RLC). Based on results of the radio resources management algorithms executed in the RRM entity, it can be necessary to optimize the radio interface operation and therefore change some of the parameters of layers 1 and 2 on the fly. Some reconfiguration procedures of the radio channels are specified in the RRC layer and in the physical layer in order to apply these new parameters. The current implementation requires to close and to open these channels again, but in the future it will be possible to modify these parameters in a transparent way, by means of the reconfiguration procedures. Some parameters such as the transmission power can already be adjusted based on the result of the radio measures.

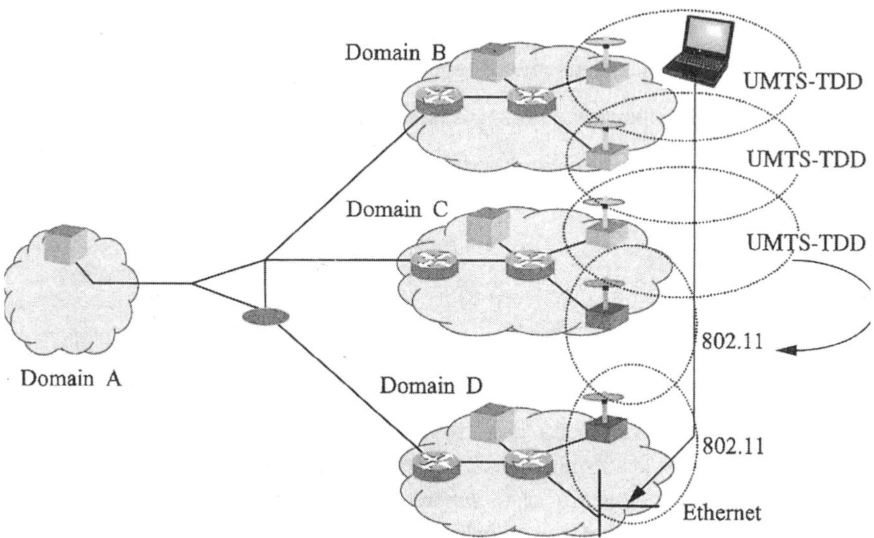

Figure 4.8. *Mobility in a "pure-IP" environment*

The RRM entity can also decide to change the radio context by moving the data traffic towards a WLAN interface when one is available (and vice versa). We then execute a *handover* or mobility procedure directly at the IP level. The direct interconnection with the IP protocol enables an easy reconfiguration of this context, without the creation of tunnels and the intervention of intermediate protocols, as it is shown in Figure 4.8. We plan to make this mobility even more flexible through the addition of a *soft handover* mechanism at the IP level making it possible to duplicate the data flow on the two radio interfaces (UMTS and WLAN) throughout the transfer. This mechanism, which is implemented with WLAN interfaces, must be carried on carried on the UMTS-TDD platform soon.

4.5.3. *Software reconfigurability*

A third reconfiguration possibility of the platform is currently being designed and tested as part of the IST E2R [IST] project.

In fact, the reconfiguration of the protocol stack, which is total or partial, can be potentially executed during the mobile operation by downloading a software component containing either some source code or some precompiled modules. A typical scenario is that of the software update when the terminal is started up. A user switches on his terminal which executes an initialization phase during which a software, operating as an update client, is automatically executed and activates a default communication with a server application. After verification of the user identity, this user then has access to a list of updates that are compatible with his system and his current software versions. The user chooses a new functionality, a problem correction or any other platform improvement. The selected software component is then downloaded in the platform and possibly authenticated by its signature. It is then executed and replaces a part of the existing software by the source or object code coming from the server. Depending on the level of the modification executed, the terminal may be rebooted so that the update is recognized. It is also possible to replace the whole software of the radio platform in this way or to trigger the reconfiguration from a specific component of network monitoring if this is considered necessary. This could allow the implementation and experimentation in real-life conditions of mass upgrade algorithms for mobile terminals.

4.6. Conclusion

In this chapter we have introduced a software radio platform implementing an UMTS-TDD type waveform. The entire signal processing, as well as the radio access protocol stack are implemented in software on some commercially available processor running with RTLinux. The hardware part is thus limited to the digital/analog conversions, to the radio frequency handling and to the power components. This approach enables a total reconfiguration of the waveforms considered and of the associated processing (encoding, equalizing, and radio resources management, etc.).

4.7. Bibliography

[23.060] 3GPP TS 23.060, *Technical Specification Group Services and System Aspects; General Packet Radio Service (GPRS), Service description.*

[25.201] 3GPP TS 25.201, *Physical Layer – General Description.*

[25.301] 3GPP TS 25.301, *Radio Interface Protocol Architecture.*

[25.331] 3GPP TS 25.331, *RRC Protocol Specification.*

[3GP] 3GPP website, http://www.3gpp.org.

[EUR] www.wireless3G4Free.com.

[IEE] IEEE working groups – IEEE 802.xx, http://grouper.ieee.org/groups/802.

[INT] http://www.intel.com.

[IP] IPv6 Forum website, http://www.ipv6forum.org.

[IST] Projects IST FP6, http://www.cordis.lu/ist/projects/projects.htm.

[IST 04] WETTERWALD M., BONNET C., GAUTHIER L., MORET Y., KNOPP R., NUSSBAUM D., MELIN E., "An original adaptation of the UMTS protocols for a direct interconnection with IPv6", *13th IST Mobile and Wireless Communications Summit 2004*, 27-30 June, Lyon, 2004.

[LIN] Linux website, http://www.linux.org.

[MES] http://www.projectmesa.org.

[MIP] Mobile IP Charter, http://www.ietf.cnri.reston.va.us/html.charters/mobileip-charter.html.

[MOB] Project IST Moby Dick, http://www.ist-mobydick.org.

[RNR] http://www.telecom.gouv.fr/rnrt.

[RTA] RTAI website, http://www.rtai.org.

[RTL] FSMlabs website, http://www.fsmlabs.com.

[SDR 03] BONNET C., CALLEWAERT H., GAUTHIER L., KNOPP R., MENOUNI H.A., MORET Y., NUSSBAUM D., RACUNICA I., WETTERWALD M., "Open-source experimental B3G networks based on software-radio technology", *SDR'2003, Software digital radio*, Orlando, November 2003.

[SIP] SIP Charter, http://www.ietf.org/html.charters/sip-charter.html.

Chapter 5

Iterative Approach for Hardware Reconfigurability: The Rake Receiver

5.1. Introduction

In second generation mobile telecommunications systems, the radio parameters, except for the power control, are usually set statically so that the system can operate while taking into account a degradation of the circuit quality. For example, the modulation parameters and the coding scheme are initialized at the beginning of the communication in order to take the "worst case" into account and they do not change anymore afterward. This static method of parameterization specific to the HDR (*hardware defined radio*) principle is suboptimal, from a service point of view, in order to guarantee "quality", if there is a modification of the channel, as well as from an electronic point of view, where the consumption of the portable receiver is unnecessarily increased. The new standards and protocols offer a better flexibility for the management of parameters and thus the possibility of improving the quality of the communication [HOL 01]. In order to statically or dynamically adapt to the environment, mobile terminals incorporate more and more modes and communication standards. The consequence for electronic designers is a great increase of the complexity and new challenges such as finding compromises for:

– integration of a flexible, multi-standard, multi-mode system;

– minimum production cost;

Chapter written by Ioannis KRIKIDIS, Lírida NAVINER and Jean-Luc DANGER.

– possible quick launch on the market;

– minimum power consumption.

The desired flexibility of telecommunication systems can be inspired by the *software radio* principle, in which the signal is ideally digitally processed immediately after its reception in the antenna. Even though it is utopian to literally apply this principle to the letter, especially because of the limitations of analog-digital converters (see Chapter 7), it is interesting to study the way in which the flexibility of the digital part can be managed as well as possible, while taking compromises into account. A 100% software digital processing is flexible in essence, but even by taking into account the technological improvements of the DSPs (*digital signal processors*), it is clearly suboptimal in terms of consumption and performance. A minimum of material processing must still be necessarily considered.

The concept of flexibility can be viewed under two complementary aspects: "adaptiveness" and the "reconfigurability" [POL 03]. A system is adaptive if the digital values of its parameters can change dynamically. For example, the modification of the number of steps in the LMS (*least mean square*) algorithm illustrated in Figure 5.1 corresponds to the modification of a digital value and thus meets the adaptiveness criterion.

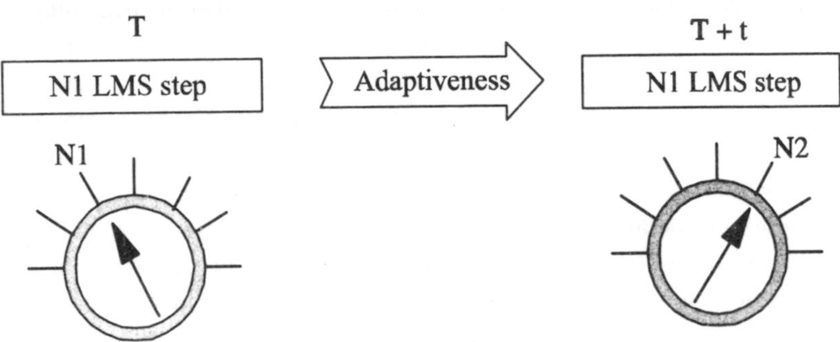

Figure 5.1. *Example of adaptiveness: modification of the step of the LMS algorithm*

The reconfigurability implies the possibility of modifying dynamically a function by a change that is not quantifiable, i.e. without sufficient representation by a digital value. For example, the shift from a convolutional coding scheme to a turbo-code, as illustrated in Figure 5.2, cannot be represented by a change of the parameter's value, hence we deal with reconfigurability.

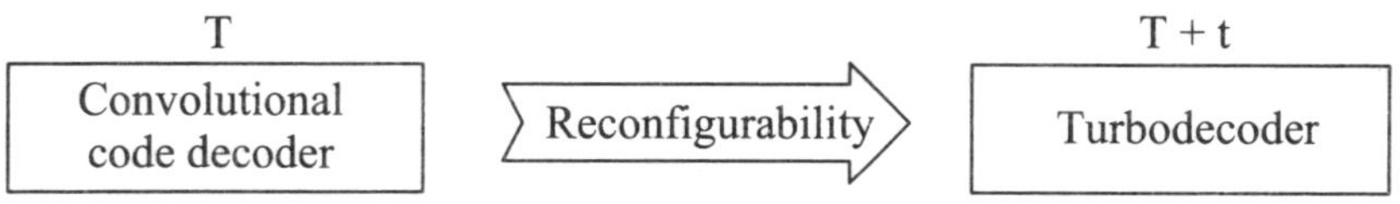

Figure 5.2. *Example of reconfigurability: code change*

A software design of a flexible system is therefore naturally adaptable and reconfigurable. However, the search for the best compromise (cost, development time, performance, consumption) requires the common use of software and optimized hardware structures that are specific to the "SoC" (*system on a chip*) structures. The question is to know whether the hardware reconfigurability makes sense in the field of multi-standard communication systems; most of the designers can select the use of pre-designed material units that correspond to each standard. Reconfigurability consists here of switching from one unit to another. If the advantage of this solution is to decrease the development time, it is certainly suboptimal in silicon surface and consumption. Another approach that we are going to develop consists of studying the common or close structures between various communication modes and standards. The following sections discuss a point of view on the hardware reconfigurability with significant examples based on a *rake* operator of a DS-CDMA receiver.

5.2. Concept of hardware reconfigurability

Hardware reconfigurability consists of a circuit that can achieve various functions at different times. A microprocessor is an example of reconfigurable circuit by considering that its instruction set is a set of configurable functions. However, its serial and generic operation generates a decrease of the overall performance and an increase of the consumption that does not always make it possible to meet the specified physical constraints. Concerning this latter issue, an "ASIC" (*application specific integrated circuit*) type dedicated electronic circuit is recommended but is naturally not configurable. In the "SoC" on chip systems we observe the complementarity of the processors and of the dedicated material; that acts as a "coprocessor" accelerating the calculations, or as an execution unit that is specific to an "ASIP" (*application specific instruction set processor*) processor. A programmable material circuit FPGA (*field programmable gate array*) is potentially a reconfigurable circuit if it is possible to reprogram it on the fly in order to change its functionality. Its calculation refinement as well its generic character makes it a very costly component and a very "consuming" one. Hardware reconfigurability can therefore only make sense by using an adapted and optimized reconfigurable structure within an ASIC in order to optimize the constraints: performance, consumption and cost.

5.2.1. *The "multiplexing" approach*

In this approach, the idea is to simply switch the pre-existing hardware functions. This method takes advantage of the pre-designed units and enables the quick design of an ASIC in order to have a multi-standard system. However, a big part of the hardware cannot be used. The additional cost of the surface generated by the block concatenation makes us think that a configurable structure is less expensive, consumes less and has a greater flexibility.

5.2.2. *The "pagination" approach*

In this approach, the electronic circuit is of FPGA type and its function changes according to the reconfiguration. This approach used in [SRI 00] is also known under the term of paginated material. It makes it possible to minimize the material but remains critical on fine "grain" FPGA structures that are identical to the commercial ones because the FPGA reprogramming time is often too slow to meet the performance constraints of the application. To take advantage of this approach, it is necessary to design a FPGA type matrix structure dedicated to a range of applications, where the calculation grain is more "dense" but more performing in terms of reconfiguration time and consumption.

5.2.3. *The "factorization" approach*

This approach benefits from operators that are common to a group of functions to be configured. We consider a set of functions $S = (f_1, f_2 \ldots, f_n)$ that have a minimum number of common operators, which can be expressed by $f_i \cap f_j \neq \varnothing$ for each i,j [GRA 00]. By using the same material basis formed by all the operators, the reconfiguration consists of changing function while choosing the suitable operators with the associated connections for each function f_i. Figure 5.3 illustrates the operation of this approach where there is the platform formed of M operators and a

SPV supervision component in charge of generating the control signals to manage the operators and the connections.

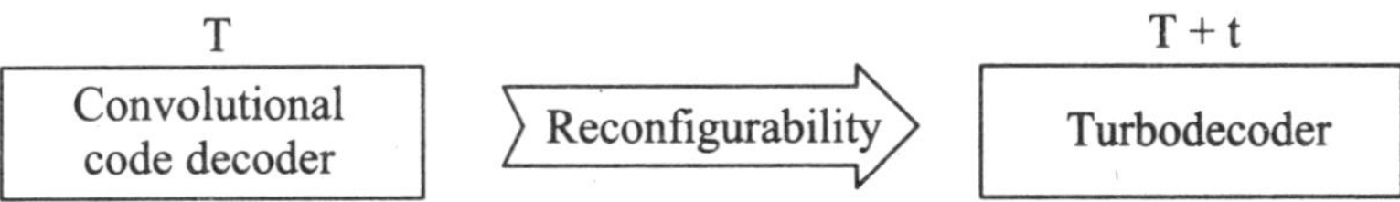

Figure 5.3. *Platform configured for two different functions*

5.2.4. *The "iteration" approach*

This approach based on the "factorization" approach attempts to take advantage of the algorithms compatible with the serialization of calculations by having operators whose "granularity" is finer and in which the function can loop several times. The communication systems use numerous functions that can easily be "serialized" or whose nature is iterative; this is the case of the turbo-code and LDPC (*low density parity check codes*) decoders but also the equalization, which will be studied in the following chapters. Figure 5.4 illustrates the approach proposed with base operators which are smaller and submitted to several iterations.

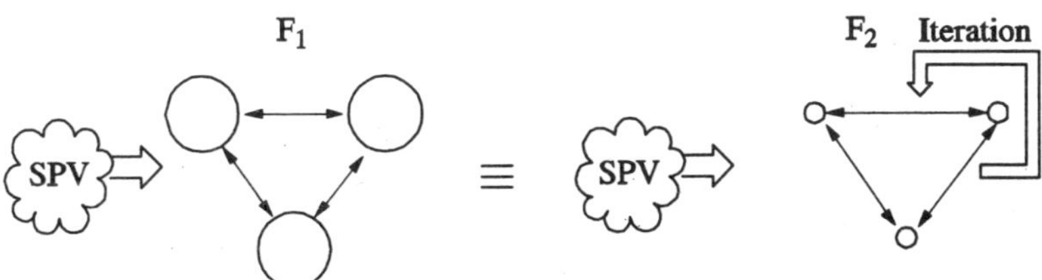

Figure 5.4. *Platform with fine grain operators for the serialization*

The proposed approach makes it possible to decrease the size of the hardware platform and authorizes an adjustment of the algorithm's quality by acting on the number of iterations. The dynamic consumption of this type of serial structure is usually greater than that of the "factorization" structure for the same calculation power. However, the static consumption that becomes significant in the CMOS $< 0.1\ \mu m$ technologies is clearly smaller due to the reduced size of the calculation grain. For some functions, all the operators can be unused, but the low granularity of the operators minimizes the suboptimal use of the material. The number of iterations can also be limited by the time allocated for the processing. However, by considering that an iteration can only last for a clock cycle with a *pipeline* architecture, the ratio between the clock frequency (>> 200Mmots/s) and the binary rates (from a few Mbits/s to a few dozens MB/s) offers a number of iterations that is sufficient for numerous telecommunication functions. The examples described further on illustrate the interest in this method.

5.3. Example 1: reconfigurable rake receiver with an "interference canceller"

This example falls in the context of a DS-CDMA (*direct sequence code division multiple access*) receiver used in the "3G" third generation portable telephony standards. This algorithm illustrates an example of hardware reconfigurability with the "iteration" approach.

5.3.1. *Formulation of the problem*

In a 3G system, it is possible to change the communication rate by acting on the spreading factor that is specific to the DS-CDMA principle. To provide a very high rate service, a low spreading factor (2 or 4) is needed. At these rates, there is a high probability to have only one user or a user who is not disturbed by other users who have higher spreading factors. In this case, the inter-users interferences (MAI: *multi-access interference*) can be negligible. However, the interference between "IPI" (*interpath interference*) symbols connected to a channel with numerous paths is often very high and can significantly degrade the performance of a system [KRI 04a].

By considering a transmission model of a base station toward a terminal (*downlink*) in a multipath Rayleigh fading channel, the baseband signal received can be written:

$$r(t) = \sum_{m=1}^{M} \sum_{l=1}^{L} \alpha_l(m) b(m) s(t - mT - \tau_l) + n(t) \qquad [5.1]$$

where r(t) is the signal received, M is the length of the observation slot in symbols, L is the number of paths, $\alpha_l(m)$ is a complex number that represents the fading factor, $b(m) \in \{\pm 1 \pm 1j\}$ is the QPSK symbol transmitted, $\tau_l \in [0\ \ T]$ is the propagation delay of the path l, T is the duration of the symbol and n(t) is the additional Gaussian white noise. In equation [5.1], signal s(t) is spread with the SF spreading factor on SF "chips" according to formula [5.2]:

$$s(t) = \sum_{n=0}^{SF-1} c(n)h(t - nT_c) \qquad [5.2]$$

where T_c is the duration of a chip, c(n) is the value of the chip n and h(t) is the chip pulse of duration T_c.

The *rake* receiver is in charge of "despreading" the signal s(t) on the L traveled paths and then to recombine the result of the L paths. By assuming a logical detection, a perfect knowledge of the channel and a recombination of the rake "fingers" by means of the MRC (*maximum ratio combining*) algorithm, the signal coming from the *rake* receiver can be expressed by:

$$\tilde{b}(m) = \sum_{l=1}^{L} \alpha_l^*(m) \int_{(m-1)T+\tau_l}^{mT+\tau_l} r(t)s(t - \tau_l)dt = W + I + N \qquad [5.3]$$

where W, I and N represent respectively the demodulated signal, an interference component and the noise. They are expressed by the following equations:

$$W = \sum_{l=1}^{L} |\alpha_l(m)|^2 b(m)T \qquad [5.4]$$

$$\begin{aligned} I &= \sum_{l=1}^{L} \sum_{\substack{q=1 \\ q \neq l}}^{L} \alpha_l(m)\alpha_q^*(m) \int_{(m-1)T+\tau_l}^{mT+\tau_l} b(t - \tau_l)s(t - \tau_l)s(t - \tau_q)dt = \\ &= \sum_{l=1}^{L} \sum_{\substack{q=1 \\ q \neq l}}^{L} \alpha_l(m)\alpha_q^*(m) \begin{bmatrix} b(m - 1 - \delta_{l,q})R(\tau_{l,q} - \delta_{l,q}T) \\ + b(m - \delta_{l,q})\hat{R}(\tau_{l,q} - \delta_{l,q}T) \end{bmatrix} \end{aligned} \qquad [5.5]$$

$$N = \sum_{l=0}^{L} \alpha_l^*(m) \int_{(m-1)T+\tau_l}^{mT+\tau_l} n(t)s(t-\tau_l)dt \qquad [5.6]$$

where $\delta_{l,q} = \lfloor \tau_l - \tau_q / T \rfloor$, $\tau_{l,q} = \tau_l - \tau_q$, $R(\tau) = \int_0^{\tau} s(t-\tau)s(t)dt$ and $\hat{R}(\tau) = \int_{\tau}^{T} s(t-\tau)s(t)dt$ for $0 \le \tau \le T$ are Pursley's partial equations [PUR 77].

The term I represents an IPI component interference. This term comes from the fact that it is impossible to design spectrum spreading codes that are perfectly orthogonal for all the time shifts. The "despreading", which corresponds to an autocorrelation function, therefore generates a non-zero result when the signal is shifted, as it is expressed in equation [5.7].

$$R'(\tau) = R(\tau) + \hat{R}(\tau) = \begin{cases} 1 \text{ for } \tau = 0 \\ -\dfrac{1}{SF} \text{ for } \tau \neq 0 \end{cases} \qquad [5.7]$$

For high rates, i.e. when SF is low, the rake receiver is more strongly degraded by the IPI. An interference cancellation system can be used to decrease the negative impact of this interference.

5.3.2. *Proposed algorithm*

The interference cancellation (IC) is based on the knowledge of the noise generators that form the other users in the case of the MAI or the other symbols in the case of the IPI. The principle is to reproduce the interference so that it is subtracted afterwards, in order to generate information where the interference noise is decreased [DIV 98; HUI 98]. It is also possible to reiterate the process by using the signal already processed during a first iteration as the input signal. In our example, this technique is applied to suppress the IPI with only one user. Figure 5.5 shows the functional segmentation of the receiver.

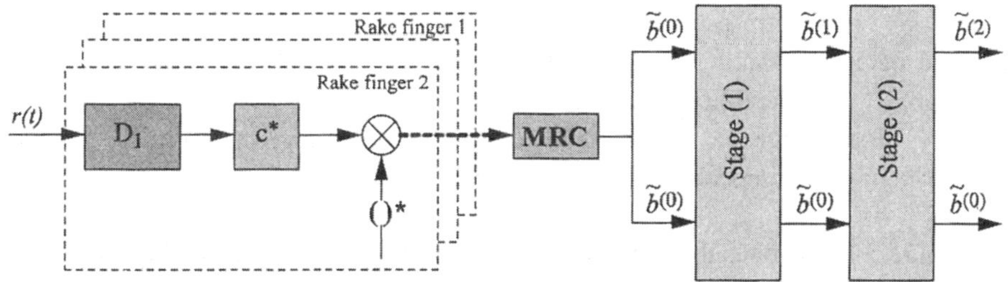

Figure 5.5. *Rake functions and multi-stage interference canceller*

After the *rake* receiver, several stages to cancel the interference are used. The structure of the stage i is illustrated in Figure 5.6.

The first operation of an interference suppressor stage consists of estimating the *rake* receiver output signal of the i – 1 stage. A decision function is used to this end. It can be expressed by:

$$\hat{b}^{(i-1)}(m) = f_{dec}(\tilde{b}^{(i-1)}(m)) \quad [5.8]$$

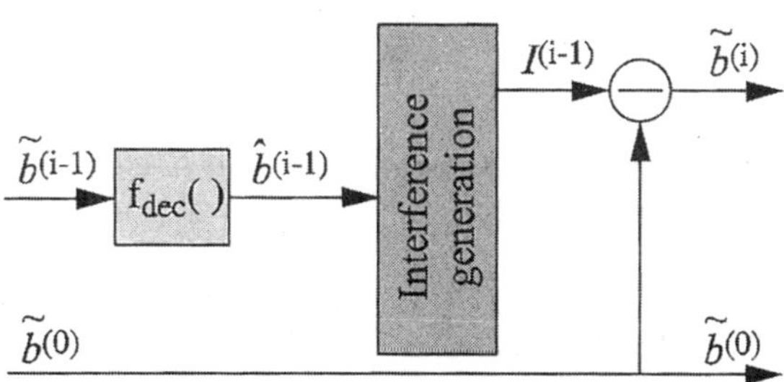

Figure 5.6. *Structure of the i stage*

Then the signal is remodulated as if it were a transmitter, i.e. the spreading operation is carried out and the delays identical to those of the term I of the IPI are generated. Then the signal is processed as in a rake receiver in order to regenerate the IPI. At each stage output, the interference generated is subtracted to the signal:

$$\tilde{b}^{(i)}(m) = \tilde{b}^{(0)}(m) - I^{(i-1)}(m) \quad [5.9]$$

where $I^{(i-1)}(m)$ corresponds to the IPI reconstructed by the stage i for the symbol m. The decision function $\hat{b}^{(i-1)}$ can be on one bit (hard decision) or more (soft decision). It is preferable to use a mixed decision function in order to combine the rapidity of convergence of the hard decision, if it is correct, with the highest guarantee of convergence of the soft decision [ZHA 03].

Figure 5.7 illustrates the decision function applied to the real and imaginary parts of the signal. When the signal is strong, i.e. above a certain threshold c, the hard decision is selected; otherwise a soft decision is used to avoid the propagation of the error.

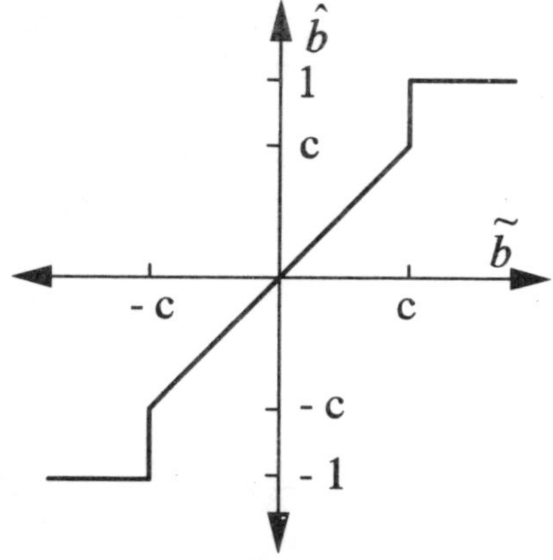

Figure 5.7. *Decision function $f_{dec}()$ with threshold c*

5.3.3. *Evaluation of performance*

Simulation results in terms of bit error rate have been obtained considering an environment and the following parameters of the UMTS standard:

– Rayleigh fading channel;

– L = 3;

– power per path = [0 0 0]dB;

– and delays τ = [0 3 6]T_c.

Figure 5.8 shows the bit error rate obtained for a SF = 4 spreading factor and 5 stages for the interference cancellation. In these conditions, the algorithm proposed is better than the (BLE-MMSE) and the (BLE-ZF) algorithm [KLE 96] whose results are also represented in Figure 5.8.

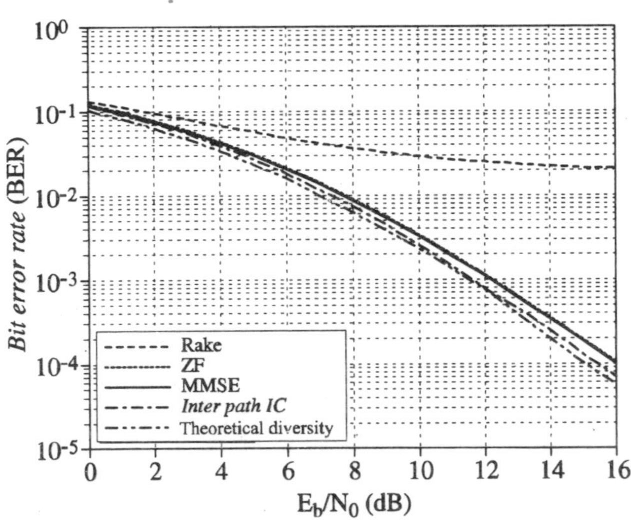

Figure 5.8. *Evaluation of the performance, evaluation of the interference suppressor*

5.3.4. *Reconfigurable architecture*

This algorithm can be achieved by benefiting from the "iteration" type reconfigurability and has a calculation structure that is formed of operators used or not, following the use in *rake* mode or in IC mode. There are as many L iterations as the number of fingers in the *rake* mode. In IC mode, the number of iterations corresponds to the number of stages of the algorithm previously described, which is multiplied by the number of the interferences per symbol = L × (L – 1) according to equation [5.5].

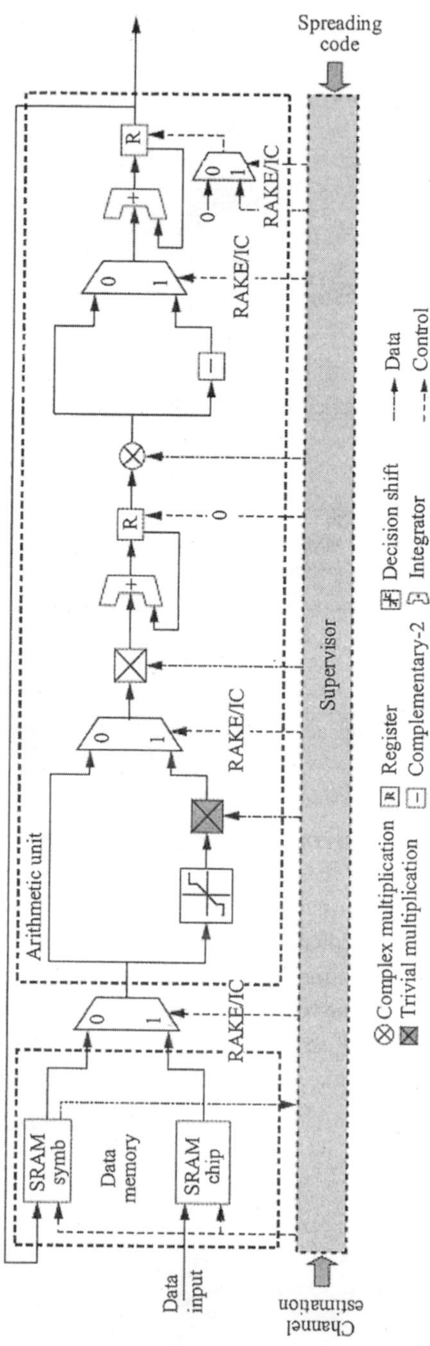

Figure 5.9. *Reconfigurable architecture*

A detailed diagram of the architecture is given in Figure 5.9. It consists of three main units: data memory, arithmetic unit and reconfiguration supervisor.

5.3.4.1. *The data memory*

The interference cancellation algorithm works on a block of data transmitted during a slot that corresponds to the temporal quantum of a packet of data of the DS-CDMA standards. The data memory consists of a static RAM (SRAM) that has two types of access: an access to receive the samples of the signal and another one to carry out the rake + IC processing of the data from the previous slot. The processing time corresponds to a slot and the two accesses overlap, as is illustrated in Figure 5.10, in order to start the processing as soon as possible and to decrease the memory size.

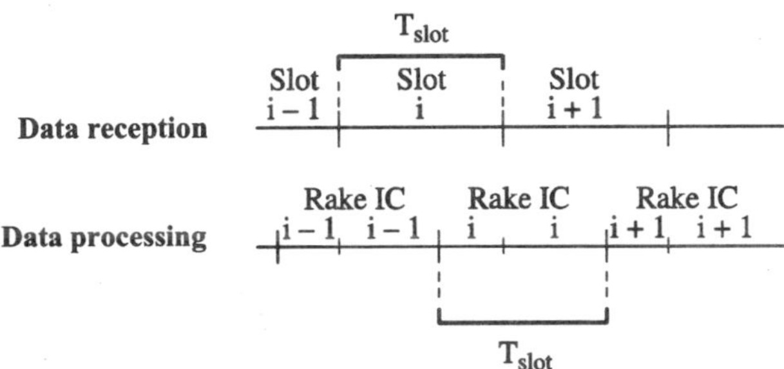

Figure 5.10. *Timing diagram of the data flow during reception and processing*

5.3.4.2. *The arithmetic unit*

In order to enable reconfigurability, this unit has operators that are common to the *rake* and IC functions as well as operators that are specific to each function. Figures 5.11 and 5.12 represent respectively the data path for the two configurations.

In the *rake* configuration, the data come from the SRAM sample memory and go directly to the "despreading" unit, that corresponds to an integrator, in order to generate the symbols. These are then multiplied by complex coefficients of the estimated channel. A final integrator achieves the MRC combination of the various calculated *rake* fingers.

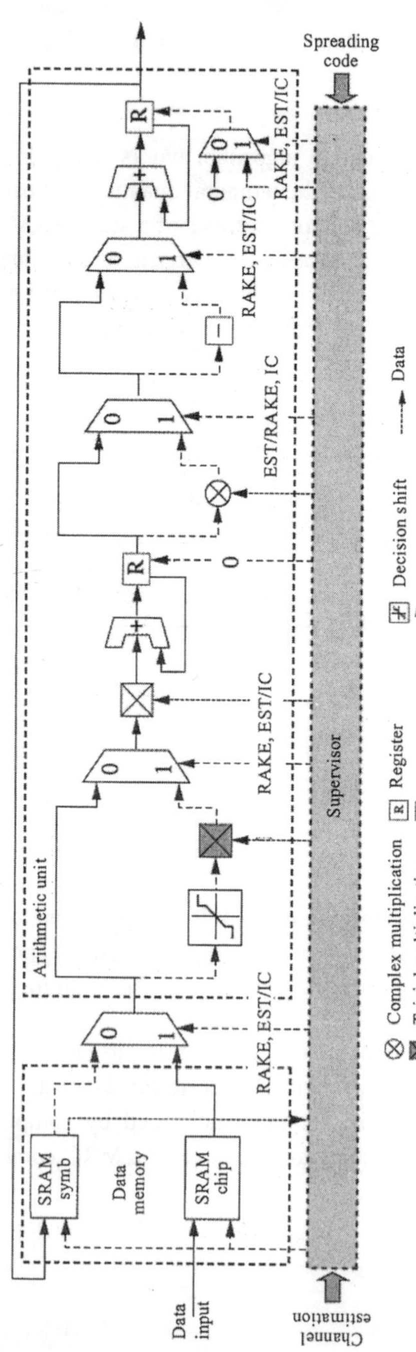

Figure 5.11. *Configuration in rake mode*

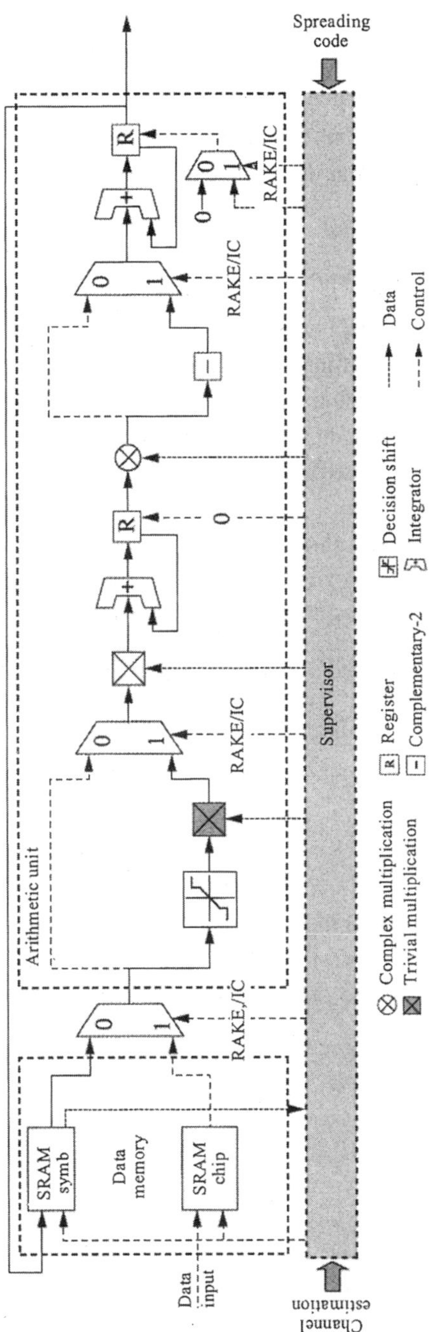

Figure 5.12. *Configuration in IC mode*

In an “interference canceller” (IC) mode, the input data are the symbols coming from the rake or the previous stage. These symbols are then submitted to the decision function and spread, which differentiates this rake function, then they are spread again with a delay $\tau_{l,q}$, which is generated by the supervisor, as expressed in equation [5.5]. After multiplication by the channel coefficients, the symbol is then complemented and accumulated in order to generate the overall interference term to be subtracted from the symbol.

5.3.4.3. *The configuration supervisor*

The SPV supervisor provides the control signals to the arithmetic unit and to the memory. It is in charge of reconfiguring the arithmetic unit in order to shift from a mode to another. It carries out the translation of the estimated times by the channel estimator, in data address, in order to read the samples that correspond to the estimated delays from the SRAM. The supervisor can have a dynamic reconfiguration algorithm in order to find the best parameters (L, n) according to the channel and the number of processing cycles that we have.

By considering a synchronic and pipeline operator that operates at the frequency F_{clk}, the calculation time of an iteration is $\frac{SF}{F_{clk}}$ because it is necessary to integrate the SF chips in the symbol in order to “unspread”. The processing time is therefore $[L+V.L.(L-1)]\frac{SF}{F_{clk}}$ by considering L paths and V stages for the interference canceller.

The number of possible iterations is limited by the symbol time $\frac{SF}{F_{chip}}$. The values of L and V are therefore constrained according to the equation:

$$L+V.L.(L-1)<\frac{F_{clk}}{F_{chip}} \qquad [5.10]$$

It is possible to obtain $\frac{F_{clk}}{F_{chip}}$ of the order of 100 by considering the CMOS $< 0.1\ \mu$ technologies, which would give the following possible values of $L = 5$ and $V = 4$.

In the case where the channel paths have very different powers, the supervisor can choose a partial cancellation algorithm, which results in a decrease of the term L.(L – 1) and consequently of the calculation consumption.

5.4. Example 2: an interference canceller based on realistic channel estimation

In the first algorithm, we considered a perfect knowledge of the channel in order to implement the logical detection and the process of interference cancellation. In this example, the algorithm takes into account the channel estimation operation whose quality has a direct impact on the performance of a terminal. The proposed algorithm is also adapted to the DS-CDMA broadband connections. Based on a multi-stage approach, it aims at suppressing the interferences of the driver channel and the data channel.

5.4.1. *Formulation of the problem*

The descending communication model considered is based on a single-user DS-CDMA system on a Rayleigh type fading channel and with the channel estimation per driver. The transmitted signal can be expressed by:

$$e(t) = P_0 s_0(t) + P_1 b(t) s_1(t) \qquad [5.11]$$

where $s_0(t)$, $s_1(t)$ indicate signature sequences for the driver $(n = 0)$ and the data $(n = 1)$; P_0 and P_1 are, respectively, the amplitudes of the driver and the data. It must be mentioned that no data symbol is present on the driver channel.

The signal received can thus be written as [5.12], where n(t) is a Gaussian white noise with bilateral power spectral density $N_0/2$:

$$r(t) = \sum_{l=1}^{L} \alpha_l(t)\left[P_0 s_0(t-\tau_l) + P_1 b(t-\tau_l) s_1(t-\tau_l)\right] + n(t) \qquad [5.12]$$

This is the case of a non-coded transmission, with delays τ_l that are perfectly known. In order to estimate the channel coefficients that are necessary for a consistent demodulation, the signal received is “unspread” from the spreading code of the driver channel. The fading channel complex coefficient estimated for the jth path and the mth symbol transmitted is thus given by [5.13]:

$$\tilde{\alpha}_j = 1/SF \cdot P_0 \int_{(m-1)T+\tau_j}^{mT+\tau_j} r(t) s_0(t-\tau_j) dt = \alpha_j + I_{1j} + I'_{1j} + N_{1j} \qquad [5.13]$$

where I_{1j} and I'_{1j} are respectively the interferences with the data path and with the driver channel paths, and N_{1j} is the noise component. In addition, each interference component can be rewritten as:

$$I_{1j} = P_1 / SF \cdot P_0 \sum_{\substack{l=1 \\ l \neq j}}^{L} \alpha_l \int_{(m-1)T+\tau_l}^{mT+\tau_l} b(t-\tau_l) s_1(t-\tau_l) s_0(t-\tau_j) dt \qquad [5.14]$$

$$I'_{1j} = 1/SF \sum_{\substack{l=1 \\ l \neq j}}^{L} \alpha_l \int_{(m-1)T+\tau_l}^{mT+\tau_l} s_0(t-\tau_l) s_0(t-\tau_j) dt \qquad [5.15]$$

$$N_{1j} = 1/NP_0 \int_{(m-1)T+\tau_l}^{mT+\tau_l} n(t) s_0(t-\tau_j) dt \qquad [5.16]$$

The quality of the channel estimation based on the driver symbols is decreased by the multipath interference terms I_{1j} and I'_{1j} that introduce an estimation error.

The demodulation operation is similar to that of channel estimation, except for the fact that the despreading is carried out with the data spreading code. The rake receiver is the conventional demodulation mode. By supposing the MRC (*maximum ratio combining*) algorithm and the L available demodulation fingers that are associates with the L channel paths, the variable decision of the mth data symbol is:

$$\tilde{b}(m) = 1/SF \cdot P_1 \sum_{j=1}^{L} \tilde{\alpha}_j^* \int_{(m-1)T+\tau_j}^{mT+\tau_j} r(t) s_1(t-\tau_j) dt = D + I_2 + I'_2 + N_2 \qquad [5.17]$$

where D is the component of the signal of generated diversity I_2 and I_2' represent respectively the interferences between data paths and those between data paths and the driver, and N_2 is the noise component.

The various components at the correlator's output are written as:

$$D = b(m)\sum_{j=1}^{L} \alpha_j \tilde{\alpha}_j^* \qquad [5.18]$$

$$I_2 = 1/SF \sum_{j=1}^{L} \sum_{\substack{l=1 \\ l \neq j}}^{L} \tilde{\alpha}_j^* \alpha_l \int_{(m-1)T+\tau_j}^{mT+\tau_j} b(t-\tau_j) s_1(t-\tau_j) s_1(t-\tau_l) dt \qquad [5.19]$$

$$I_2' = P_0 / SF \cdot P_1 \sum_{j=1}^{L} \sum_{\substack{l=1 \\ l \neq j}}^{L} \tilde{\alpha}_j^* \alpha_l \int_{(m-1)T+\tau_j}^{mT+\tau_j} s_0(t-\tau_j) s_1(t-\tau_l) dt \qquad [5.20]$$

$$N_2 = 1/SF \cdot P_1 \sum_{j=1}^{L} \tilde{\alpha}_j^* \int_{(m-1)T+\tau_j}^{mT+\tau_j} n(t) s_1(t-\tau_l) dt \qquad [5.21]$$

These equations show that the signal detection is penalized by the IPI (I_2, I_2') and is also sensitive to the estimation errors ($\tilde{\alpha}$). For very high rates, the IPI and the estimation error generated can lead to a degradation of the system performance rather than a gain of the multipath diversity.

5.4.2. *Proposed algorithm*

The general structure of the proposed multi-stage interference canceller is shown in Figure 5.13. The basic idea is to reproduce the terms of interference, and then to subtract them in order to generate "cleaned" channel estimates and data. The successive application of this process should lead to a big improvement of the system performance.

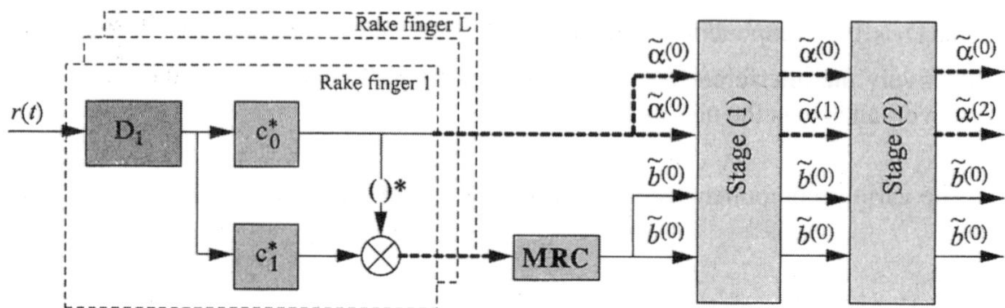

Figure 5.13. *IC interpath with channel estimate*

In the initial stage, the receiver – which works in conventional mode – demodulates and unspreads the signal received. The initial outputs of the correlator ($\tilde{b}^{(0)}(m)$) and the channel estimator $\left(\tilde{\alpha}^{(0)} = \left[\tilde{\alpha}_1^{(0)} \quad \dots \quad \tilde{\alpha}_L^{(0)}\right]\right)$ are used for each stage of the multi-stage canceller. The objective of the proposed canceller is to attenuate these interferences.

The structure of the ith interference stage is presented in Figure 5.14. The main operation is the estimate of the signal transmitted. In order to achieve this, a hard decision is made at the output of the i – 1 stage. This can be expressed as [5.22]:

$$\hat{b}^{(i-1)}(m) = \text{sgn}[\tilde{b}^{(i-1)}(m)] \qquad [5.22]$$

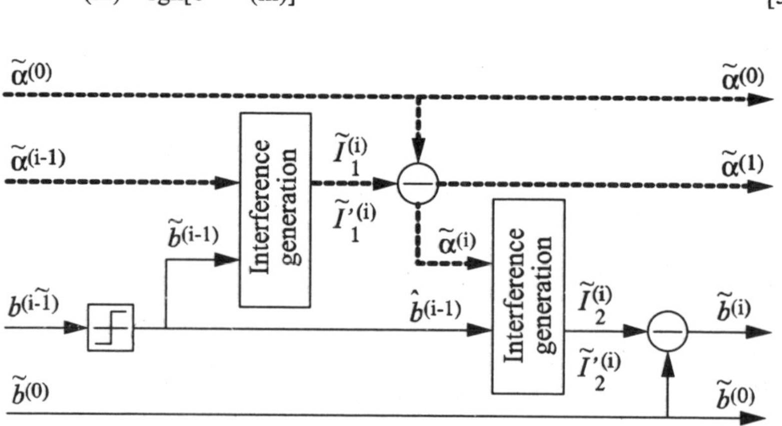

Figure 5.14. *Structure of the i-th cancellation stage*

This estimate is combined with the channel estimate of the stage output $i-1$ in order to reproduce the interference terms included in the initial channel estimation. If $\tilde{I}_1^{(i)} = \left[\tilde{I}_{1,1}^{(i)} \ \ldots \ \tilde{I}_{1,L}^{(i)}\right]$ and $\tilde{I}_1'^{(i)} = \left[\tilde{I}_{1,1}'^{(i)} \ \ldots \ \tilde{I}_{1,L}'^{(i)}\right]$ are the duplicates of the interference terms, the updated estimation channel is [5.23]:

$$\tilde{\alpha}^{(i)} = \tilde{\alpha}^{(0)} - \tilde{I}_1^{(i)} - \tilde{I}_1'^{(i)} \qquad [5.23]$$

$\tilde{\alpha}^{(i)}$ and $\hat{b}^{(i-1)}(m)$ are used in order to reproduce the duplicates of the interference terms presented at the output of the initial correlation. By assuming that $\tilde{I}_2^{(i)}$ and $\tilde{I}_2'^{(i)}$ are the generated interference terms, the new correlation output is:

$$\tilde{b}^{(i)}(m) = \tilde{b}^{(0)}(m) - \tilde{I}_2^{(i)} - \tilde{I}_2'^{(i)} \qquad [5.24]$$

It is necessary to have good data and channel estimations so that this iterative diagram can operate properly. Otherwise, the performance will be degraded by a generated propagation error.

5.4.3. *Evaluation of the performance*

The simulation environment is based on the specifications of the FDD-UMTS standard in the descending communication. The radio channel consists of $L = 2$ independent Rayleigh multipaths, whose average power is Pt = [0 0]dB and whose average delay is τ = [0 2]Tc. The power ratio between driver channel and data channel is equal to 6.5 dB. Finally, the rake receiver used considers $L = 2$ fingers.

MSE (mean square error) comparisons between the conventional estimate (V = 0) and the method proposed for various numbers of stages are presented in Figure 5.15a. We notice that the estimate channel improves when the number of stages increases. After a certain number of iterations (V > 6), the method converges to the 13 dB value.

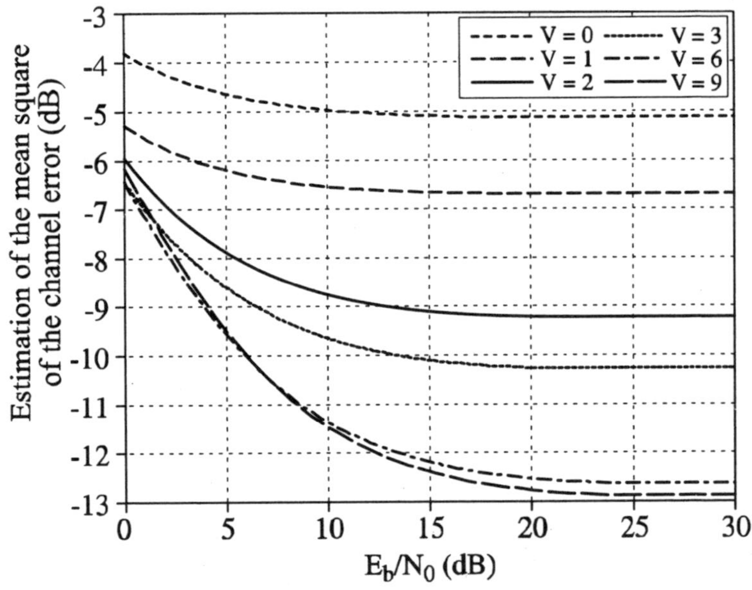

Figure 5.15a. *MSE performance of the estimation diagram proposed versus the traditional diagram (V = 0)*

Figure 5.15b provides the bit error rate (BER) of the diagram proposed for various numbers of stages. The curves confirm the observations previously made and attest to the accuracy of the method proposed. We can also see that the propagation error on the MSE curves does not lead to a degradation of the BER performance.

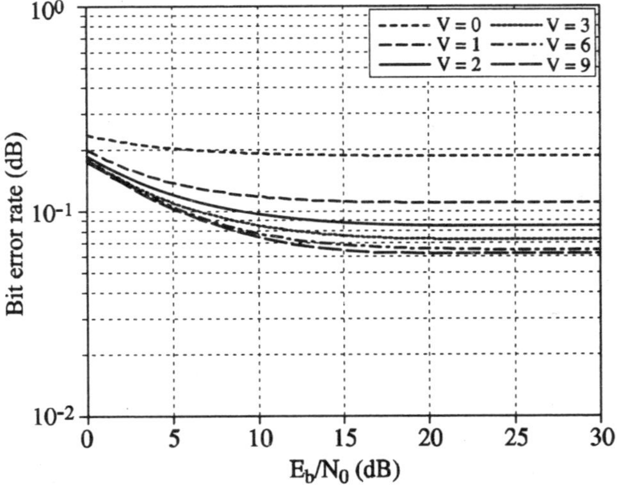

Figure 5.15b. *BER performance of the estimate diagram proposed versus the traditional diagram ($V = 0$)*

5.4.4. *Reconfigurable architecture*

As in example 1, the generation process for the various types of interference presents many similarities with the rake demodulation and channel estimate operations. All these functions are iterative and it is possible to benefit from the "iteration" type reconfiguration approach. The structure of the receiver is very close to the one presented for example 1. It contains the three calculation units, memory and configuration supervision. The operating diagram of the arithmetic block is presented in Figure 5.16. The supervisor is in charge of generating the signals from A1 to A6 for each mode.

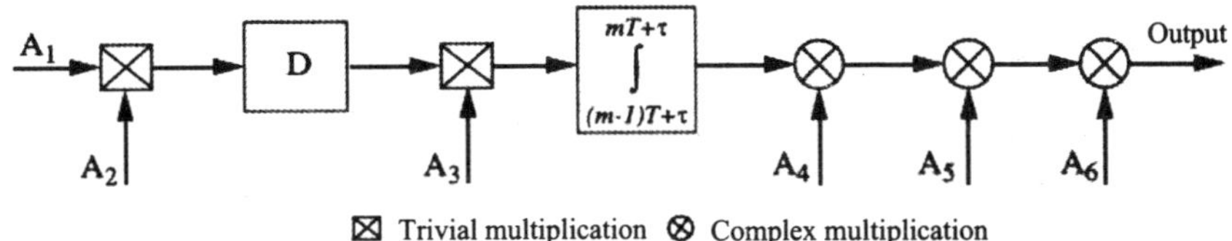

Figure 5.16. *Base processing unit for the IC with channel estimate*

Table 5.1 provides the type of signal and the type of coefficient to be generated for each mode.

Mode	Number of iterations	A_1	A_2	D	A3	A4	A5	A6	Output
Rake	L	r(t)	1	0	$s_1(t-\tau_j)$	1	$\tilde{\alpha}_j^*$	$1/SF \cdot P_1$	Finger j
Channel	L	r(t)	1	0	$s_0(t-\tau_j)$	1	1	$1/SF \cdot P_0$	$\tilde{\alpha}_j$
IPI data (channel estimate)	V.L(L – 1)	$\hat{b}(t)$	$s_1(t)$	τ_l	$s_0(t-\tau_j)$	$\tilde{\alpha}_l$	1	P_1/SFP_0	$\tilde{I}_{1j,l}$
IPI channel (channel estimate)	V.L(L – 1)	1	$s_0(t)$	τ_l	$s_0(t-\tau_j)$	$\tilde{\alpha}_l$	1	$1/SF$	$\tilde{I}'_{1j,l}$
IPI data (detection)	V.L(L – 1)	$\hat{b}(t)$	$s_1(t)$	τ_j	$s_1(t-\tau_l)$	$\tilde{\alpha}_l$	$\tilde{\alpha}_j^*$	$1/SF$	$\tilde{I}_{2j,l}$
IPI channel (detection)	V.L(L – 1)	1	$s_0(t)$	τ_j	$s_1(t-\tau_l)$	$\tilde{\alpha}_l$	$\tilde{\alpha}_j^*$	$P_0/SF \cdot P_1$	$\tilde{I}'_{2j,l}$

Table 5.1. *The various configurations*

The arithmetic unit presented in the first example can be used with a few small modifications to support the "EST" channel estimate configuration. Figure 5.17 presents the data path for the new configuration. As we can see, the data path is a simplification of the *rake* configuration. There are not the complex multiplication and the return to the MRC unit. The channel estimate consists of a "despreading" with the spreading sequence that corresponds to the driver channel and to an integration of SF chips. This process makes it possible to refine the channel coefficient corresponding to the path and the symbol considered at each iteration.

The total processing time is $[2.L+4.V.L.(L-1)]\frac{SF}{F_{clk}}$ by considering L paths and V stages for the interference canceller with channel estimation. The term 2L illustrates the time for the rake operations for the data channel on the one hand and the driver channel estimation on the other hand. The term 4V.L.(L – 1) comes from the four ISI interference terms of the driver channel and the data channel as indicated in Table 5.1.

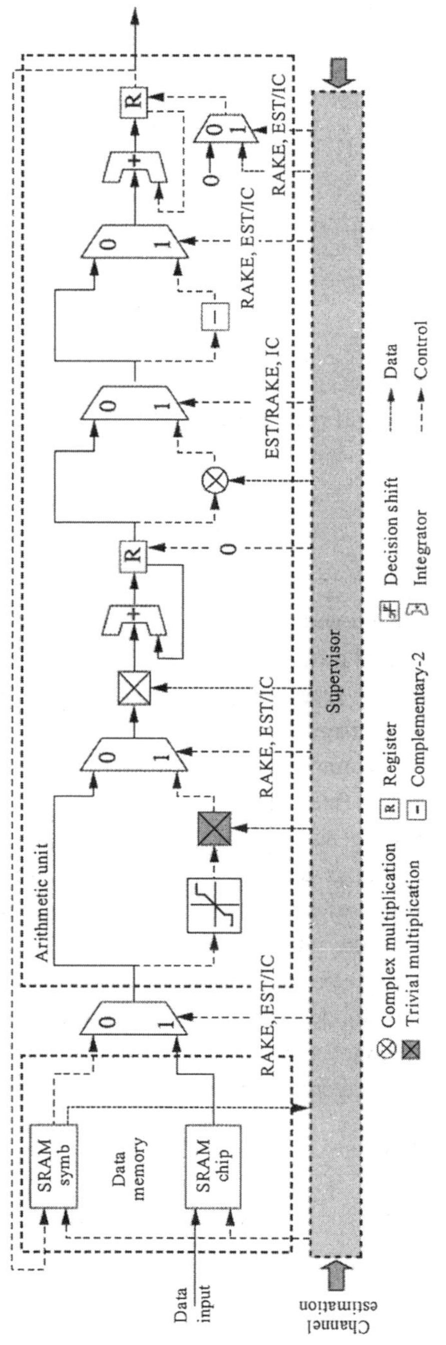

Figure 5.17. *Configuration in EST mode*

The number of possible iterations is limited by the symbol time $\frac{SF}{F_{chip}}$. The values of L and V are therefore constrained according to the equation:

$$2.L + 4.V.L.(L-1) < \frac{F_{clk}}{F_{chip}} \quad [5.25]$$

It is possible to obtain $\frac{F_{clk}}{F_{chip}}$ of the order of 100 by considering the CMOS 0.9 μ, technologies, which would give the following possible values of $L = 2$ and $V = 10$.

5.5. Conclusion

In this chapter, we saw that it was possible to apply the hardware reconfiguration approaches adapted to the signal processing for digital communications. The generic "multiplexing" or "pagination" type approaches can be optimized by using the calculations that are almost similar and iterative. The architectures can therefore be limited to operators that are common to all functions ("factorization" approach). Moreover, they can have a very fine calculation "grain" in order to operate iterative calculations that are specific to numerous communication algorithms ("iteration" approach). Examples have been given regarding the equalization in a rake receiver where the algorithms in charge of decreasing the interferences and to estimate the channel lie on a refinement of the information at each iteration. The interest of this approach is on the one hand a greater algorithmic flexibility by acting on the number of iterations and configuration modes, and on the other hand a cost of silicon, which is a significant static consumption for nanotechnologies, and a lower development time due to the small size of the architecture to be reconfigured.

5.6. Bibliography

[DIV 98] DIVSALAR D., SIMON M.K., RAPHAELI D., "Improved Parallel Interference Cancellation for CDMA", *IEEE Trans. on Communications*, vol. 46, no. 2, p. 258-268, 1998.

[GRA 00] GRAYVER E., "A Reconfigurable 8 GOP ASIC Architecture for High-Speed Data Communications", *IEEE Journal on Selected Areas in Communications*, vol. 18, no. 11, p. 2161-2171, 2000.

[HOL 01] HOLMA H., TOSKALA A., *WCDMA for UMTS-Radio Access for Third Generation Mobile Communications*, John Wiley & Sons, 2001.

[HUI 98] HUI A.L.C., LETAIEF K.B., "Successive Interference Cancellation for Multiuser Asynchronous DS/CDMA Detectors in Multipath Fading Links", *IEEE Trans. on Communications*, vol. 46, no. 3, p. 384-391, 1998.

[KLE 96] KLEIN A., KALEH G.K., BAIER P.W., "Zero forcing and minimum mean-square-error equalization for multiuser detection in code-division multiple-access channels", *IEEE Trans. Vehic. Tech.*, vol. 45, p. 276-287, 1996.

[KRI 04a] KRIKIDIS I., DANGER J.L., NAVINER L., "A Finger Configuration Algorithm for a Reconfigurable Rake Receiver", *IEEE WCNC'04*, Atlanta, USA, p. 311-315, March 2004.

[KRI 04b] KRIKIDIS I., DANGER J.L., NAVINER L., "A DS-CDMA Multi-Stage Inter Path Interference Canceller for High Bit Rates", *IEEE ISSSTA'04*, Sydney, Australia, p. 405-408, September 2004.

[POL 03] POLYDOROS A. *et al.*, "Wind-Flex: Developing a Novel Testbed for Exploring Flexible Radio Concepts in an Indoor Environment", *IEEE Comm. Magazine*, vol. 41, no. 7, p. 116-122, 2003.

[PUR 77] PURSLEY M.B., "Performance Evaluation for Phase-Coded Spread-Spectrum Multiple-Access Communication-Part I", *IEEE Trans. on Communications*, vol. 25, p. 795-799, 1977.

[SRI 00] SRIKANTESWARA S., REED J.H., ATHANAS P., BOYLE R., "A radio architecture for reconfigurable platforms", *IEEE Comm. Magazine*, vol. 38, p. 140-147, 2000.

[ZHA 03] ZHA W., BLOSTEIN S.D., "Soft-Decision Multistage Multiuser Interference Cancellation", *IEEE Trans. on Vehicular Technology*, vol. 52, no. 2, p. 380-389, 2003.

Chapter 6

Antenna Arrays and Reconfigurable MIMO Systems

6.1. Introduction

Mobile wireless communications have always presented great challenges mainly because of the particularly hostile nature of the mobile radio channel, which is affected by large scale signal fading (shadowing), by short term multipath fading and by spectral spreading due to a Doppler shift of multipath components. The dawn of digital wireless transmission technology, the dazzling growth of the wireless service user population and the tendency towards continuously higher broadband links are three factors which conspire to create outstanding technical obstacles during the continuous evolution of wireless communications. We know for example that mobile systems are especially limited by the *interference* level, which is directly proportional to the number of users.

In this context, we are also witness to a multiplication of standards, development axes and wireless services. Within wireless telephony the transition towards third generation (3G) and possibly fourth generation (4G) systems as well has started, progressively bringing a network, multi-rate and multi-service oriented operating mode, confining vocal communication and its relatively modest bandwidth to a secondary level. Let us mention among others wireless local area networks (WLANs); Wi-Fi systems, under the aegis of the IEEE[1] [IEE] 802.11 working group, operate with a significant commercial success in the 2.4 GHz and 5 GHz license free bands within

Chapter written by Sébastien ROY and Jean-Yves CHOUINARD.

1 IEEE: *Institute of Electrical and Electronics Engineers.*

most parts of the world. HiperLAN2 networks, under the aegis of ETSI[2] [ETS], are cast in the same mould. There are also the metropolitan networks (WMAN[3]) and those for wide area coverage (WWAN[4]). This type of system is characterized mostly by a fixed access system with very large bandwidth and a long range. As an example we can mention the BRAN[5] standard created by ETSI and Wi-Max systems, the latter being in their initial phase of commercial deployment, which are defined by the IEEE 802.16 standard. This last standard has the distinctive feature of covering a range of possible physical layers between 2 GHz and 66 GHz including the license-free bands of 2.4 GHz and 5 GHz recognized by many countries.

We also have to mention the nascent standard 802.20, which addresses the high mobility broadband networks (up to 250 km/h). With cells of a 15 kilometer diameter, these networks can serve high speed trains. At the other extreme, there are networks with very short range (PAN[6]) such as Bluetooth, Bluetooth 2 and Zigbee (standard 802.15.4).

In this environment where the spectrum is more and more congested, where many standards and services coexist and where signal quality is plagued by co-channel and inter-symbol interference, more and more sophisticated signal processing techniques have to be used. Space-time adaptive antenna arrays and MIMO[7] systems are such techniques, allowing the co-existence of a larger number of users with each enjoying a higher throughput without using additional spectrum.

This chapter will deal with space-time reconfigurable transceivers. In section 6.2 we address the problem of large bandwidth transmission within wireless channels by presenting the advantages provided by the reconfigurability of communication equipment. Antenna arrays have proven to be particularly efficient within wireless channels due to the use of diversity principles: they are described in section 6.3. Section 6.5 presents the concept of dynamic reconfigurability for multiple antenna systems. Section 6.6 contains application examples of reconfigurable multiple antenna systems.

6.2. Large broadband transmission and reconfigurable transceivers

6.2.1. *General context*

During the last two decades, wireless communications have had an outstanding development, overtaking even the growth rate of the computer engineering industry.

2 ETSI: *European Telecommunications Standardization Institute.*

3 WMAN: *Wireless Metropolitan Area Networks.*

4 WWAN: *Wireless Wide Area Networks.*

5 BRAN: *Broadband Radio Access Network.*

6 PAN: *Personal Area Networks.*

7 MIMO: *Multi-Input, Multi-Output.*

This relentless technical progress is motivated by a double pressure: a continuous increasing number of users along with a requirement for more and more diversified digital services which are increasingly demanding at the same time in terms of bandwidth and binary throughput.

So, it was necessary to increase on the one hand the *user-capacity* (number of users simultaneously supported by a global or local communication system) and the *throughput-capacity* (average or maximum throughput supported per user). This constitutes a double pressure on the radio spectrum, a resource which is already congested.

The great technical challenges in the wireless field come from this question of capacity to which is added, in the context of this ongoing evolution, a set of less fundamental but also crucial considerations: quality of service, interoperability between systems and standards, new and increasingly hostile propagation conditions, etc.

6.2.1.1. *Quality of service*

Along with the progressive abandoning of the traditional telephony paradigm (voice oriented and connection structured) in favor of the network paradigm (multi-service oriented and packet structured), service quality control mechanisms become indispensable. In effect, the packet network infrastructure basically depends on resource sharing. Dedicated connections are no longer possible and hence it is necessary to ensure that communications with specific demands in terms of binary throughput (being able to be specified by the average or minimum binary throughput) and latency (e.g. conversation, video-conferencing) have their packets treated as a priority in comparison to less demanding communications. The ATM service class model constitutes the traditional paradigm in this field and the Internet IPv6[8] infrastructure is inspired by this. Nevertheless, the issue is getting more and more complicated (and the stipulated mechanisms must consequently evolve), especially within the wireless area of the systems with a multiplication of standards, superposition of systems within certain bands (particularly the license-free bands), UWB[9] radio, multiplication of services and binary throughput (multi-rate systems) and generalized spectral congestion.

6.2.1.2. *Interoperability*

The time when a great number of communication systems could be operated in the comfort of a relative isolation is no longer here. In order to unite local wireless networks into vast metropolitan networks, long range wireless links are used. The license-free bands, which are requested more and more (802.11, 802.16, wireless telephone), have the characteristic of being totally anarchical. In fact, except for certain Government constraints on transmission power and the direct gain of the antennae, everything

8 IPv6: *Internet Protocol, version 6.*

9 UWB: *Ultra-Wide Band.*

is allowed. It follows that these systems will, in the future, have to put up with an unpredictable propagation environment where interferences will have unknown modulation types and traffic patterns. It is safe to assume that the systems operating within these bands will have to be more and more flexible and adaptive (i.e. *cognitive*) in order to remain reliable in this anarchical environment which is to become more and more congested.

6.2.1.3. *Propagation conditions*

In order to counter the congestion problems and to allow high binary throughput, one of the options considered is to assign higher frequency bands to the systems (e.g. millimetric bands) which have relatively little congestion and therefore make it possible to allocate contiguous spectrum blocks of considerable bandwidth, thus facilitating the creation of broadband wireless links. However, propagation within such bands is often unknown and very hostile, wave behavior being quasi-optical. In fact, studies have shown that within the 28–30 GHz band, an unobstructed line of sight is practically indispensable in order to obtain a reliable communication, a fact which considerably limits the system coverage and range. Besides, such a short wavelength considerably amplifies the multipath and Doppler phenomena, requiring more sophisticated, effective and costly adaptive mechanisms to block them. Rain attenuation is an equally serious problem within millimetric bands.

Furthermore, and this is valid for any band, we know that the most hostile feature of wireless channels is caused by mobility. This refers to Doppler effects, which, combined with multipath propagation, generate spreading in frequency and time-selective fading. High binary throughputs exacerbate these problems by creating (or amplifying) the inter-symbol interference, making carrier and symbol-time recovery more difficult and the use of great constellations (which would allow a higher throughput without using additional bandwidth) practically impossible for all ends. Even the so-called fixed systems such as 802.16 or BRAN are subject to these Doppler effects. In fact, it was proven that the multipath fading within the millimetric bands (where these wide range fixed systems usually operate being used as a backbone for possible metropolitan networks) are as severe as – actually more so – in the case of mobile systems [HON 97]. This is because of scatterer movements, amplified by the shortness of the wavelength. From the point of view of the millimetric waves, the smallest objects, even the tree leaves waving in the wind, become moving scatterers. In addition, practically all surfaces become *rough* from an electromagnetic point of view [BEC 63], thus generating a very great propagation range. Furthermore, the oscillatory movements of buildings and structures supporting antennae generate important Doppler effects.

Taking into account all these technical challenges, the trends seem to indicate that future wireless communications will be characterized by a coexistence of standards, an increasing diversity of services with different needs in terms of binary throughput and quality of service, a constant growth of the number of subscribers involving a growth of the number of links that the systems will have to simultaneously support and a growth of the bandwidth used by each user.

In addition, the majority of wireless development axes display a convergence towards a unifying paradigm. This paradigm, which we shall call *broadband wireless access*, has the following significant features:

– binary throughput exceeding firstly 10 Mb/s, then possibly 100 Mb/s, and even 200 Mb/s;

– a global access, universal and transparent access, via intelligent reconfigurable terminals;

– anywhere and anytime availability;

– the support of various degrees of mobility.

It is easy to verify that the main axes converge in this direction. In fact, cellular telephony, with the possible dawn of the 4G (predicted for 2006-2020), is devoted to permanently abandoning the telephonic paradigm in favor of the network paradigm, and the transition promises high binary throughput. Wireless networking already offered nominal throughout of 54 Mb/s (802.11a/g), but since 2006 it promised a throughput of 100-200 Mb/s (802.11n) through space-time encoding. In addition, local wireless networks evolve in order to accept mobility and *handover* (intercellular relays). Regarding metropolitan/wireless networks, which are under the aegis of the 802.16 working group, a nomadic variant is being developed (mobility support in 802.16e).

6.2.2. *Reconfigurable radios*

In the context of the evolution of wireless communications presented above, it is clear that terminals and base stations will have to show a continuously increasing degree of flexibility and intelligence. The concept of software-defined radio, which has nevertheless been used for some time, thus becomes more and more relevant. Originally, software-defined radio consisted of digitizing the entire useful band of the system, followed by a reception chain (channel selection, frequency translation, filtering, demodulation, detection, etc.) entirely implemented as software on a digital signal processor (DSP). This offers many advantages such as a certain protection against obsolescence (because the material can support new algorithms based on a software update) and a faster development cycle.

However, to date, the concept of software-defined radio has not been an outstanding success in the market. In fact, there are many practical obstacles. For example, even if the main part of the communication chain benefits from the flexibility of a software implementation, the RF part (from the antenna to the sampler) remains dedicated. This raises cost constraints and a certain rigidity concerning the system band, the width and the position of individual channels. The problem is increased for antenna array transceivers because they have to duplicate the entire reception chain for all antennae. It is therefore better to minimize the importance of RF sections by sampling as close as possible to the antennae. This nevertheless requires samplers-and-hold and higher speed analog-to-digital converters because it is necessary to cover a wider total band around a higher central frequency (in the absence of downconversion to baseband). However, the samplers and converters already constitute the bottleneck of the software-defined radio chain given the manufacturing difficulty associated with high sampling rates. These components are therefore expensive and demanding in terms of power, which limits their use in portable and mobile applications, whose battery life is a chief criterion for competitiveness.

In addition, the creation of software-defined radio based on digital signal processors involves a relatively expensive infrastructure (high performance processor, memory, etc.) which is not always fast enough to perform all the necessary operations, especially at high throughput. Therefore, we speak nowadays more about *reconfigurable radios*, which use reconfigurable circuits like FPGAs, possibly in combination with DSP. This allows more flexibility and more rapidity, possibly avoiding the entire infrastructure associated with high performance DSP, but at the cost of a longer development cycle. This is why DSPs, possibly less expensive and less bulky because of reduced performance requirements, continue to play a role because they make it possible to easily implement the control logic (state machine) as well as operations considered to be difficult to design on FPGAs (e.g. the matrix inversion), while the latter are almost exclusively dedicated to operations which require the highest performance (e.g. FFT, translation in digital frequency, etc.).

Since hardware achievements offer an undeniable performance advantage over software, FPGAs are very likely to be quickly incorporated in most common digital equipment, including communication terminals (mobile telephones, PDAs, etc.). This will provide great flexibility making it possible to redefine at will the functionality of a device. It will therefore be possible to modify the internal circuitry in order to adjust to recent technological advances, which will bring about a certain degree of protection against obsolescence.

It is a concept whose time has come, as the following quote from Verkest [VER 03] testifies:

> By year-end 2005, US consumers will have trashed some 130 million cell phones and another mountain of old PDAs, MP3 players and game consoles. We could, of course, build bigger landfills [...]. But here is a much better idea: building a wireless multimedia device whose hardware and software can be easily altered or upgraded so it never becomes obsolete.

Despite the unquestionable potential of this development process, certain practical considerations are paramount, especially in the short term. One of them is the great power consumption by most FPGAs in comparison to dedicated circuits. There is also a speed penalty for taking the FPGA route over equivalent dedicated circuits. In the communication field, where mobile applications and battery life are paramount, the ideal is a hybrid solution. In fact we can consider the production of a competitive and low consumption dedicated circuit, capable of performing all critical operations (which are fundamentally of a generic nature) quickly, but incorporating a portion of FPGA material, thus providing a degree of flexibility to the organization of these operations in order to be in conformity with, for example, a given standard.

Dynamic reconfiguration

However, the use of FPGAs raises the intriguing possibility of performing *partial reconfiguration* or *dynamic reconfiguration*. In fact, some of these circuits can be reconfigured during operation, one part at a time, while the rest of the circuit continues to operate normally.

This capacity of partial/dynamic reconfiguration is still little used in practice, but is the subject of much research work. It needs the use of FPGAs which supports it or DFPGAs[10]. Ideally, these kinds of FPGAs will make it possible to partition the die area in an arbitrary manner, thus establishing the parts to be reconfigured and those which must function normally. Moreover, they would have a fine granularity (the base logic cell would be small) for more flexibility and their configuration time would be short. If certain DFPGAs manufactured in the past had these ideal features (e.g. Xilinx XC6200, a futuristic device strongly appreciated in research, but discontinued due to lack of sufficient market), the currently available DFPGAs (such as Xilinx, Virtex, Virtex-Spartan-2 and Spartan 3 families, Atmel AT6000, AT40K and AT94K families and Stratix family) are much more limited, imposing a coarse partition structure with a relatively long reconfiguration time. Furthermore, the design flow for a dynamic reconfigurable system is relatively complex and it seems to get even more complicated because the encoding format of bit streams defining the configuration is invariably a sealed secret of the manufacturer. Hence, these bit streams cannot be directly controlled and partitioned; it is instead necessary to pass through utilities, which, for the moment, remain basic, even if they are progressing rapidly.

Let us consider for example Xilinx FPGA families. Partitioning is done only in logic block columns [XIL 04b]. Thus, a reconfigurable module always occupies the full height of the device. For example, the XC2V4000 device of the Virtex-II family consists of 72 columns by 84 rows of logic blocks, while the XC2V8000, the largest device of the same family, constitutes a matrix of 112×104 logic blocks (CLB[11])

10 DFPGA: *Dynamically Reconfigurable FPGA.*

11 CLB: *Configurable Logic Block.*

[XIL 04a]. It is clear that these partition rules impose severe constraints in terms of position and routing and, inevitably, a smaller utilization of the device.

The fastest configuration method for these devices is Xilinx *Select MAP* method which makes it possible to transfer one byte at a time at a maximum frequency of 50 MHz. Thus, in order to completely transfer the 15,659,936 configuration bits of the XC2V4000, $15,659,936/8\ (50\ \text{MHz})^{-1} = 39.15$ ms are necessary, which is not negligible. In the same manner, the initial configuration of the XC2V8000, which involves 26, 194, 208 configuration bits, needs 65.5 ms. The dynamic reconfiguration of a module requires a time fraction of the complete configuration based on the number of occupied columns. Since this delay is in milliseconds in these complex devices and thus corresponds to millions of clock seconds, frequent reconfigurations should be avoided to maintain any complexity advantage linked to dynamic reconfiguration.

However, there is a potentially faster reconfiguration technique: differential reconfiguration [XIL 04b]. Based on the fact that the multiple versions of a reconfigurable module show only minor changes between them, it is possible to transfer only the *differences*, which constitutes a configuration sequence much shorter than that describing the entire circuit. However, the generation of similar configuration sequences is delicate and requires a "manual" intervention at a very low level within the design flow.

Taking into account the partition constraints, the reconfiguration delays and the complications in the design flow, a simple approach is recommended in order to exploit dynamic reconfiguration in reconfigurable radios. This approach is based on the following principles:

– it is necessary to minimize the number of reconfigurable modules, ideally to have only one in order to simplify the design flow;

– the dynamic and static parts have to be as independent as possible, so that they can be totally decoupled during the reconfiguration phase;

– the decoupling between the static and dynamic parts has to be performed based on the principles of *temporal multiplexing*, i.e. it is necessary to isolate and regroup the functionalities that are not required at the same time, which are in fact temporarily *orthogonal* and to achieve them within the dynamic part.

In light of these various considerations, it is clear that the implementation of reconfigurable antenna arrays involves a duplication and a complexity which can be excessive, especially next to the terminal. It is therefore relevant to develop less complex algorithms and to use dynamic reconfiguration when this is adequate.

6.3. Space-time processing and MIMO systems

The electromagnetic spectrum is a finite resource and is already heavily congested. It is generally accepted that the use of antenna arrays (also given by the term *spatial diversity*) at one, at the other or at both ends of a wireless connection is without doubt the best long term solution to manage current or potential congestion problems. In fact, the use of an *adaptive antenna array* at the base stations in cellular networks can bring an increase in the number of wireless connections simultaneously supportable. An additional evolutionary threshold is crossed if we consider the use of multiple antennae at both ends (base station and terminal). This kind of connection is called a *MIMO system* and it was mathematically demonstrated (by virtue of Shannon capacity) that such systems can provide an increase by an order of magnitude of the binary throughput in a point-to-point connection with no additional consumption of the spectrum. A significant part of this potential can be obtained in practice by an adequate use of processing techniques and space-time coding.

6.3.1. *Modeling of the wireless channel*

Wireless telecommunications are invariably affected by multipath propagation which causes random signal strength fluctuations. The channel is referred to as *slow fading* when the duration of symbols transmitted by source T_s is longer than the coherence time, T_{coh}, of the transmission channel. The coherence time depends on the channel conditions, such as receiver speed in personal or mobile communications. On the contrary, when $T_s < T_{\text{coh}}$ the channel is referred to as *fast fading*.

The coherence band of a channel indicates the maximum bandwidth beyond which the spectrum components become affected differently by the propagation mechanisms. Thus, if the bandwidth B_s of the signal is wide compared to the width of the coherence band of the channel B_{coh} then the channel is frequency selective, meaning that its spectrum components will be affected differently by multipath fading.

The signal received by the receiver $y(t)$ is affected by the impulse response of the channel $h(t,\tau)$ and additive Gaussian white noise $n(t)$:

$$y(t) = h(t,\tau) * s(t) + n(t) \qquad [6.1]$$

where $s(t)$ is the transmitted signal. If the channel is non-selective in frequency, then $y(t) = h(t) \cdot s(t) + n(t)$ and if it is invariant in time: $y(t) = h \cdot s(t) + n(t)$, the impulse response being constant.

6.3.2. *Space-time processing*

The first application of antenna arrays in civil wireless communications has been without doubt the use of *spatial diversity* [JAK 74] in order to combat signal fading caused by multipath propagation. The term "diversity" refers to the availability, at the receiver, of several copies of the desired signal, all being disturbed in a different manner by their respective propagation channel. For example, in the case of a fading channel, it is highly unlikely that deep fades should simultaneously occur on all the antennae of the array.

There are obviously other types of diversity than spatial diversity. We must mention diversity in time (corresponding to a repetition code) and diversity in frequency (transmission of the same message using two carriers sufficiently separated in frequency so that their signal losses are decorrelated) which are, however, two approaches which waste bandwidth. Finally, there is path diversity which can be operated via a *rake*[12] receiver in systems where the band is very large, for example, CDMA[13].

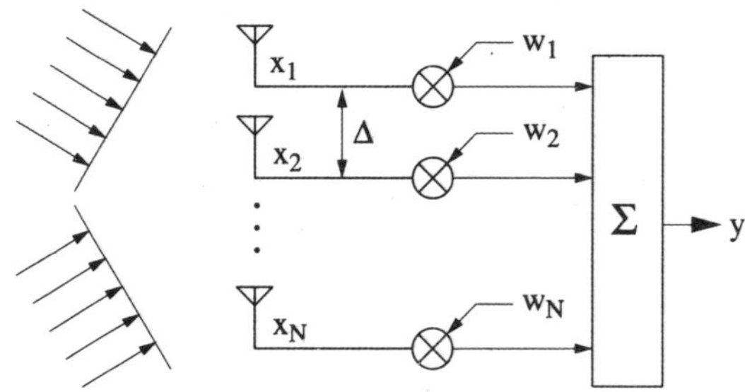

Figure 6.1. *Antenna array and linear combination processing*

The simplest type of spatial processing using diversity is doubtlessly *the selection diversity* which simply consists of always selecting the array branch having the best signal-to-noise ratio (SNR). Typical examples of architectures based on this principle will be subsequently presented. At a slightly more sophisticated level, we find the different *linear combining* diversity methods. Figure 6.1 shows the complex baseband equivalent (from where the effects of the carrier and up/downconverters are abstracted) of a linear combining antenna array. The combining operation is mathematically expressed as follows:

$$y = \boldsymbol{w}^{\mathrm{H}}\boldsymbol{x} \qquad [6.2]$$

where $\boldsymbol{w} = [w_1, \ldots, w_N]^{\mathrm{T}}$ is the weight vector, $(\cdot)^{\mathrm{H}}$ represents the conjugated transpose and $\boldsymbol{x} = [x_1, \ldots, x_N]^{\mathrm{T}}$ is the received signal. The latter is expressed as below:

$$\boldsymbol{x} = \boldsymbol{h}s + \boldsymbol{n} \qquad [6.3]$$

where $\boldsymbol{h} = [h_1, \ldots, h_N]$ is the channel vector, s is the transmitted signal, $\boldsymbol{n}$ is the noise vector and the dependence on time t is implicit in order to simplify the notation.

12 *Rake*: structure of a receiver with more branches in parallel.

13 CDMA: code division multiple access.

The received signal, consisting of the sum of different plane waves for each of the antennae, manifests itself in every point of the space where it is observed as undertaking phase and amplitude fluctuations (due to the variable phase relations between the different copies and to the destructive and constructive interference effects thus caused, because of the movement of the transmitter, receiver and/or scatterers). We can choose the complex weight in order to set all branches in phase before adding them up. This is expressed as follows:

$$\boldsymbol{w} = \left[\arg\left(h_1\right), \ldots, \arg\left(h_N\right)\right] \qquad [6.4]$$

where $\arg(\cdot)$ indicates the phase of its argument. This type of processing is called EGC.[14]

We can also weight the branches proportionally to their quality, i.e. their individual signal-to-noise ratio. In this case, the complex weight performs both this proportional weighting and branch co-phasing. We can consider that they are equal to the channel coefficients, i.e:

$$\boldsymbol{w} = \boldsymbol{h}^* \qquad [6.5]$$

This is called *maximal ratio combining* (MRC) which is optimal in the Gaussian additional white noise environment, i.e. in the absence of structured interference.

If there are other radio transmissions inside the band of interest (co-channel interference) or inside a neighbor band (adjacent channel interference) with leakage in the useful band (given filter imperfection), it is necessary to use more sophisticated combination methods. One of the most important is *optimal combining* or *minimum mean-square error (MMSC) combining*. As its name suggests, this method consists of finding the weight combination which minimizes the mean-square error (MSE) between the transmitted desired signal and the output signal of the combiner. In the same stroke, the SNR is found to be maximized. The weight vector is expressed as follows:

$$\boldsymbol{w} = \boldsymbol{R}_{\boldsymbol{xx}}^{-1}\boldsymbol{h} \qquad [6.6]$$

where:

$$\boldsymbol{R}_{\boldsymbol{xx}} = E\left[\boldsymbol{x}\boldsymbol{x}^{\mathrm{H}}\right] \qquad [6.7]$$

is the covariance matrix of the received signal and $E[\cdot]$ indicates the expectation.

Using a network of $N = M + K$ antennae with optimal combining, Winters [WIN 92] has shown that M co-channel interferers can be nulled while obtaining a Kth

14 EGC: *Equal Gain Combining*.

order diversity against fading. It should be noted that the algorithm implicitly allocates its degrees of freedom to interference nulling first, the remainder being exploited to combat fading. It is this robustness against the interferers, corresponding to the array's capability of separating signals on the same carrier, which constitutes the dominant advantage of antenna arrays within multi-user wireless communication systems. It is also this same spatial discrimination capacity which serves as a basis for the MIMO paradigm, as will be seen later.

Moreover, all these linear combination methods are also applicable to *rake* receivers which exploit path diversity. Figure 6.2 shows such a receiver, applicable to systems whose band is large enough to resolve the different paths in the channel impulse response, such as CDMA systems.

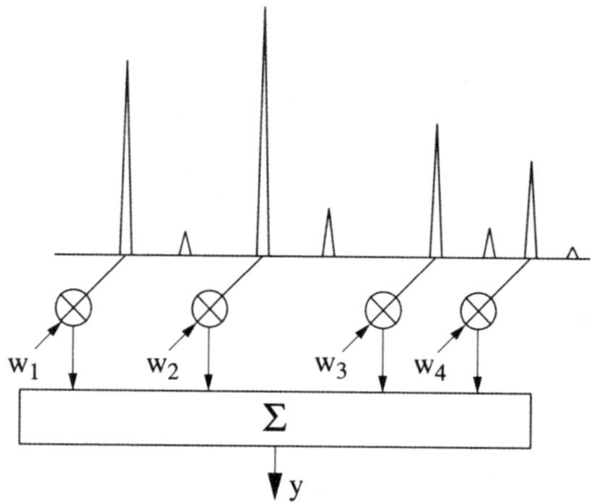

Figure 6.2. *Rake receiver principle*

The diversity methods described herein (except for the particular case of the *rake* receiver) are applicable in the case of narrowband channels, i.e. their gain and phase features do not significantly vary along the entire useful band, so that a single complex coefficient is sufficient to characterize them. In broadband channels, signal fading become frequency selective and we have to take into account a more complex impulse response in time and delay, i.e. $h(t, \tau)$. There normally implies *inter-symbol interference (IIS)*, which requires the use of *adaptive equalizers*. They constitute a form of adaptive processing in the time domain which can be performed with the same algorithms as spatial processing, for example, by minimizing the MSE. However, these same algorithms do not give the same results in both cases. Indeed, if the antenna arrays excel against co-channel interference, they offer a poor performance against inter-symbol interference. On the other hand, the equalizers manage the IIS well but have a limited nulling capacity against co-channel interference, being incapable of creating the constructive/destructive interference relations present in antenna arrays.

Hence, in the presence of these two types of interferences, it is necessary to combine equalization and adaptive antenna array, thus forming a *space-time processing receiver*. In the case of a linear combining receiver, such a system can be formed from a weight matrix corresponding to the various antennae and discrete delay to be processed. Typically, all these weights can be generated jointly by minimizing a global performance criterion, i.e. the MSE.

6.3.3. *Multiple reconfigurable antenna systems*

Figure 6.3 shows a generic model for a reconfigurable multiple antenna transceiver. For maximum flexibility, it is always recommended to push samplers as close to the antennae as possible. This is why the figure shows the samplers at IF level. However, this practice becomes expensive when there are more antennae, because it is necessary to replicate not only the RF chain, but also the analog-to-digital (A/D) and digital-to-analog (D/A) converters, which are among the most expensive and the most problematic components in terms of the power used. From the point of view of digital processing, digital front-end operations (which include frequency up/downconversion and sampling rate conversion) are equally critical because they are the only digital operations which have to be partially performed at the full rate of the samplers. Thus, this means that they are resource-intensive functions whose cost is also multiplied by the number of antennae.

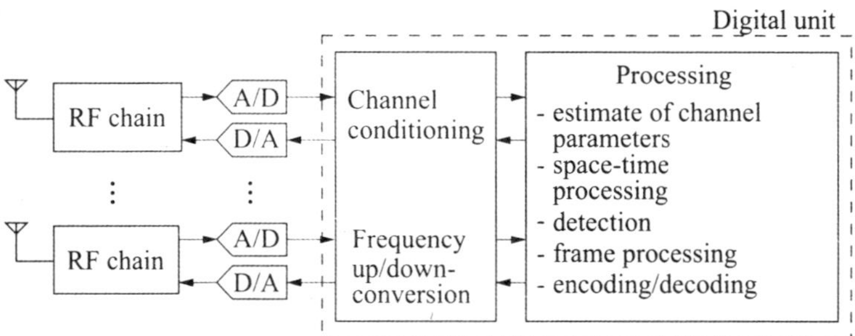

Figure 6.3. *Reconfigurable multi-antenna transceiver model*

Given the large quantity of resources used by the system presented – especially if the number of antennae is high – it is preferable to sacrifice a part of its flexibility at the space processing level by carrying it out at the RF level, i.e. ahead of the converters. We can for example imagine a selection diversity system like the one presented in Figure 6.4, which comprises only a single RF chain, a single acquisition channel and hence the subsequent processing, especially in terms of front-end operations, is considerably simplified. In addition, the systems which are subject to a certain intersymbol interference level can without problems include an equalizer (time domain processing) within the digital processing. The branch selection is done by the intervention of very simple power measuring devices which can be created at a very low cost in the form of envelope detectors based on a Schottky diode [ROY 05].

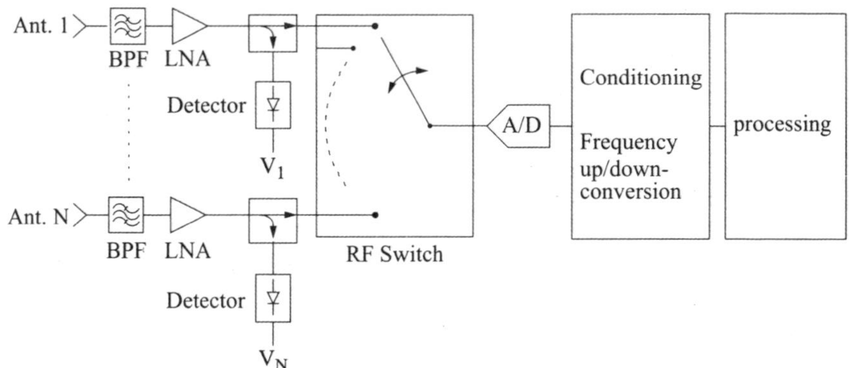

Figure 6.4. *Architecture of reconfigurable radio with selection diversity; for clarity and order the figure presents only the reception function. BPF: bandpass filter, LNA: low noise amplifier*

Obviously, complexity advantages are obtained at the expense of a fixed and moreover rudimentary space processing step. In fact, if the selection diversity supplies a certain gain against multipath fading, it is much less powerful than other space processing schemes, not very efficient against interferers and incompatible with the MIMO approach (which needs the joint processing of many antennae; see section 6.3.4).

This raises the following question: can we implement more sophisticated space processing schemes at the RF level and thus conserve the digital complexity advantage of selection diversity? We can imagine for example a form of linear space processing where each branch is balanced by a complex gain. At the RF or IF levels, this can be achieved by inserting a phase shifter and an attenuator. It is therefore sufficient to add the outputs of all branches ahead of the A/D converter. While the principle is viable, it is necessary to take into account the fact that the complex weight controlling the attenuators and phase shifters must be calculated by the digital processing unit. However, most relevant adaptation algorithms (stochastic methods with MSE minimization: LMS (*least-mean square*), DMI (*direct matrix inversion*)) need access to the signal received on *each branch* for weight calculation. If it is conceivable to develop new algorithms capable of calculating the weights based solely on the knowledge of the combined signal, little research will have been performed on this aspect, and therefore it remains an open, yet relevant research problem.

It is clear that when the number of antennae is high, the approach in Figure 6.3 is much too expensive for most applications. In this context and in light of the previous debate, how can we significantly diminish the processing complexity and the cost of RF chains without entirely sacrificing the power and flexibility associated with reconfigurable spatial processing? An interesting solution consists of selecting a *subset* of antennae to operate at the RF level as shown in Figure 6.5. This approach makes it possible to reduce the number of complete RF chains, the number of A/D and D/A conversion channels and the complexity of the digital processing, while preserving a certain number of degrees of freedom at the level of space processing and/or MIMO level. The subset selection is done using an RF switching matrix, which can be constructed cheaply based on PIN diodes [WAN 01], under the control of the digital processing unit (see Figure 6.5). A specific architecture based on this principle is presented in section 6.6.2.

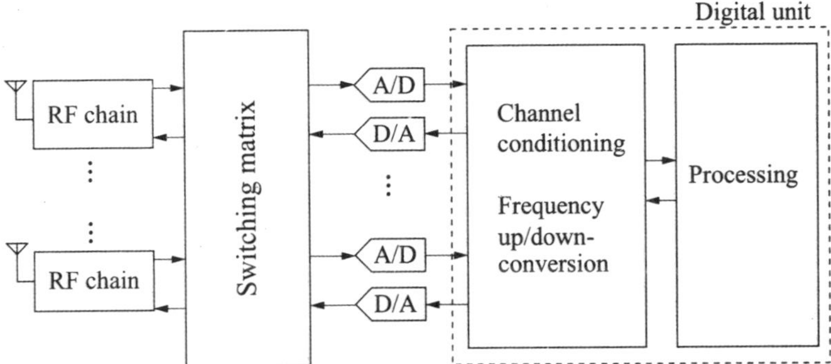

Figure 6.5. *Sub-assembly selection architecture*

6.3.4. *MIMO systems and matrix channels*

Figure 6.6 shows a telecommunication system using a radio link with two antennae at the receiver (reception diversity) and two antennae at the transmitter (transmission diversity).

The generic model of a MIMO channel is presented in Figure 6.7 with N_T transmitting antennae and N_R receiving antennae. The MIMO channel is characterized by the set of impulse responses $\{h_{i,j}(t,\tau)\}$ with $i = 1,\ldots,N_R$ and $j = 1,\ldots,N_T$, between each pair of transmitting antennae and receiving antennae:

$$\boldsymbol{H}(t,\tau) = \begin{bmatrix} h_{1,1}(t,\tau) & \cdots & h_{1,j}(t,\tau) & \cdots & h_{1,N_T}(t,\tau) \\ \vdots & \ddots & \vdots & \ddots & \vdots \\ h_{i,1}(t,\tau) & \cdots & h_{i,j}(t,\tau) & \cdots & h_{i,N_T}(t,\tau) \\ \vdots & \ddots & \vdots & \ddots & \vdots \\ h_{N_R,1}(t,\tau) & \cdots & h_{N_R,j}(t,\tau) & \cdots & h_{N_R,N_T}(t,\tau) \end{bmatrix} \qquad [6.8]$$

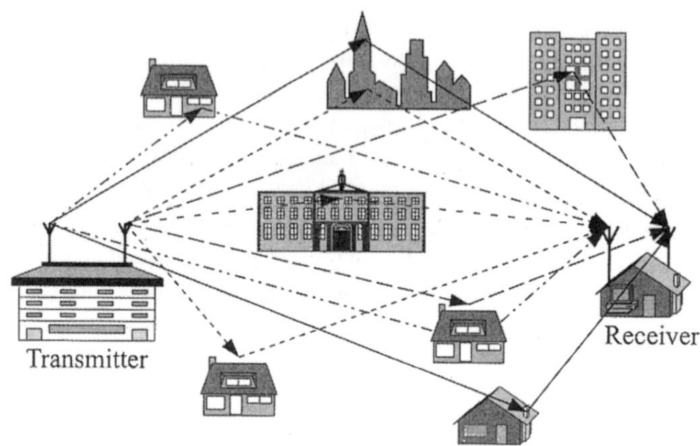

Figure 6.6. *Space-time wireless communication*

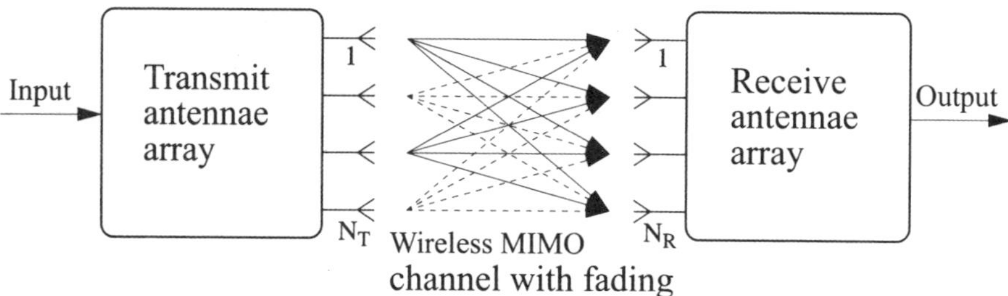

Figure 6.7. *MIMO channel model*

At each signaling interval, the transmitter sends N_T signals, $[s_1(t), \ldots, s_{N_T}(t)]$, over its N_T antennae. The signal received by the i-th receiving antenna is a mixture of N_T signals coming from the transmitting antennae and the inevitable additive noise, $n_i(t)$, at the receiving antenna:

$$y_i(t) = \sqrt{\frac{E_s}{N_T}} \sum_{j=1}^{N_T} h_{i,j}(t, \tau) * s_j(t) + n_i(t) \quad (i = 1, \ldots, N_R) \qquad [6.9]$$

The signal received by the N_R receiving antennae can be expressed in matrix form [PAU 03]:

$$\boldsymbol{y} = \sqrt{\frac{E_s}{N_T}} \boldsymbol{H} \boldsymbol{s} + \boldsymbol{n} \tag{6.10}$$

where $\boldsymbol{y}$ is of size $N_R \times 1$, $\boldsymbol{s}$ is the vector of size $N_T \times 1$ of the N_T transmitted signals which have an expectation $\mathrm{E}[\boldsymbol{s}] = \boldsymbol{0}$. The total energy available for the transmission of N_T symbols during the transmission time interval is equal to E_s and its covariance matrix is $\boldsymbol{R_{ss}} = \mathrm{E}[\boldsymbol{s}\boldsymbol{s}^{\mathrm{H}}] = (P_T/N_T)\boldsymbol{I}_{N_T}$, $\boldsymbol{I}_{N_T}$ being the identity matrix of size N_T and $\boldsymbol{s}^{\mathrm{H}}$ the matrix associated to $\boldsymbol{s}$, such that its trace is determined by the total transmission power: $\mathrm{tr}(\boldsymbol{R_{ss}}) = N_T$.

Matrix $\boldsymbol{H}$ indicates the complex gains $\{h_{i,j}(t,\tau)\}$ of the MIMO channels. Generally, they are represented by a set of independent Gaussian complex processes, with a zero mean and unitary variance, with circular symmetry, i.e. $h_{i,j}(t,\tau) \sim \mathcal{C}N(0,1)$. Supposing that the signal fades are decorrelated constitutes an analytical convenience which does not reflect reality. However, it is generally accepted that the correlation impact on performance is weak for correlation coefficients less than 0.5. The vector $\boldsymbol{n}$ is a vector of size $N_R \times 1$ of N_R noise samples at the receiving antennae. We assume the noise to be complex Gaussian additional with an expectancy $\mathrm{E}[\boldsymbol{n}\boldsymbol{n}^{\mathrm{H}}] = N_0\boldsymbol{I}_{N_R}$.

6.3.5. *Capacity of antenna arrays*

The maximum information throughput which can be reached by a MIMO system is constrained by the mutual information of the MIMO channel, i.e.:

$$I(X;Y) = \log_2 \det \left[\boldsymbol{I}_{N_R} + \frac{1}{N_T}\left(\frac{E_s}{N_0}\right)\boldsymbol{H}\boldsymbol{R_{ss}}\boldsymbol{H}^{\mathrm{H}}\right] \tag{6.11}$$

and limited by its capacity C, i.e. the maximum mutual information which can be exchanged by the system of transmitting and receiving antennae. For a MIMO system, the maximum mutual information is calculated for the group of covariance matrices $\boldsymbol{R_{ss}}$ of the transmitted vectors, but always under the constraint that its trace should be equal to the number of transmitting antennae, i.e.: $\mathrm{tr}(\boldsymbol{R_{ss}}) = N_T$, which is expressed as below:

$$\begin{aligned} C &= \max_{\mathrm{tr}(\boldsymbol{R_{ss}})=N_T} I(X;Y) \\ &= \max_{\mathrm{tr}(\boldsymbol{R_{ss}})=N_T} \log_2 \det \left[\boldsymbol{I}_{N_R} + \frac{1}{N_T}\left(\frac{E_s}{N_0}\right)\boldsymbol{H}\boldsymbol{R_{ss}}\boldsymbol{H}^{\mathrm{H}}\right] \end{aligned} \tag{6.12}$$

The capacity can be expressed in relation with the maximal binary throughput, in [bits/s], or based on the spectral efficiency, in [(bits/s)/Hz], by normalizing the MIMO system transmission bandwidth.

Often, the channel state is known only by the receiver (by estimation) and thus it is not possible to maximize the mutual information on the group of $\boldsymbol{R_{ss}}$ covariance matrices. Therefore we choose $\boldsymbol{R_{ss}} = \boldsymbol{I}_{N_T}$: the signals transmitted by the N_T antennae therefore have equal power and the mutual information is expressed as follows:

$$I(X;Y) = \log_2 \det \left[\boldsymbol{I}_{N_R} + \frac{1}{N_T}\left(\frac{E_s}{N_0}\right)\boldsymbol{H}\boldsymbol{H}^{\mathrm{H}} \right] \qquad [6.13]$$

The MIMO channel can be broken down into K orthogonal channels by a breakdown into K eigenvalues, $[\lambda_1, \ldots, \lambda_k, \ldots, \lambda_K]$, and eigenvectors, $[\nu_1, \ldots, \nu_k, \ldots, \nu_K]$ of the matrix $\boldsymbol{H}\boldsymbol{H}^{\mathrm{H}}$:

$$C = \sum_{k=1}^{K} \log_2 \left[1 + \frac{1}{N_T}\left(\frac{E_s}{N_0}\right)\lambda_k \right] \qquad [6.14]$$

The physical interpretation of the above mathematical result is as follows: the MIMO system, in a propagation environment rich in scatterers, is capable of creating through spatial discrimination a certain number of parallel and independent *virtual channels*. This is what provides the linear gain in throughput capacity with the number of antennae. However, it is neither necessary nor desirable to use all the available degrees of freedom in order to increase the throughput capacity. A portion can be reserved in order to cancel the interference factors external to the point-to-point MIMO link, or in order to increase the robustness against fading. In the latter case, we speak of *space diversity* while the throughput capacity gain deriving from the above mentioned parallel channels is named *spatial multiplexing*. We can design a MIMO system in order to obtain one or the other, or a certain compromise between these two advantages.

6.3.6. *Space-time codes*

Space-time codes are error correcting codes which introduce redundancies on two dimensions, i.e. the temporal dimension (imitating here the conventional codes) and the spatial dimension, thus using the presence of multiple transmitting antennae. It must be mentioned that a space-time code necessarily involves a MIMO system and that the opposite is not inevitably true. Figure 6.8 presents a generic model of a MIMO system with space-time encoding.

The space-time code proposed by Alamouti [ALA 98] performs the transmission of two complex symbols, s_1 and s_2, over two antennae, then retransmits in the following instant a modified version of these same two symbols. Hence, at the discreet instant $k = 1$, the transmitter transmits symbols s_1 and s_2 of the transmitting antennae 1 and 2 respectively. In the next instant, $k = 2$, the complex conjugates of these

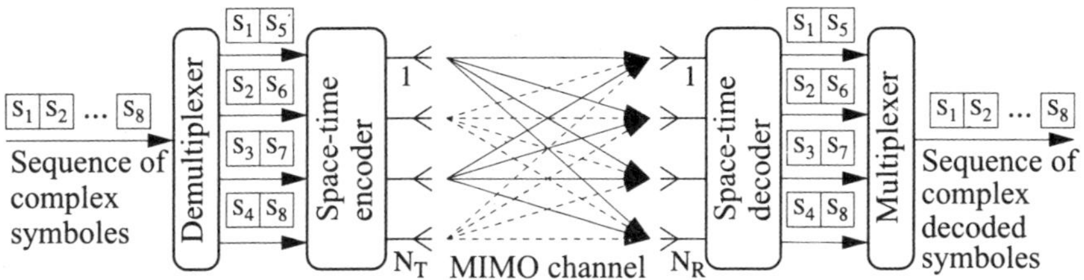

Figure 6.8. *Space-time coding principle*

two same symbols are formed, and we transmit s_2^* with the first antenna and $-s_1^*$ with the second one. The multiple antenna transmitter (here $N_T = 2$) thus forms a *transmission matrix* $\boldsymbol{S}$, defined by:

$$\boldsymbol{S} = \begin{bmatrix} s_1 & s_2 \\ s_2^* & -s_1^* \end{bmatrix} \quad [6.15]$$

Alamouti's code has a unit code rate: we transmit two complex symbols, $\boldsymbol{s} = [s_1 \quad s_2]$, using *two channel utilizations*: i.e. $\eta = 1$ complex symbol transmitted by a time unit. This is essentially a method which makes it possible to obtain a spatial diversity at the receiver when the multiple antennae are located at the transmitter.

Supposing for a moment that the communication system has only a single receiving antenna ($N_R = 1$). The signal received $\boldsymbol{r}$ is affected by signal fading in the instants $k = 1$ and $k = 2$, i.e. $\boldsymbol{h}^{\mathrm{T}} = [h_1 \quad h_2]$, $\boldsymbol{h}^{\mathrm{T}}$ being the transposed form of $\boldsymbol{h}$, as well as by the additive Gaussian noise, $\boldsymbol{n}^{\mathrm{T}} = [n_1 \quad n_2]$:

$$\boldsymbol{r} = \boldsymbol{S}\boldsymbol{h} + \boldsymbol{n} \quad [6.16]$$

$$\begin{bmatrix} r_1 \\ r_2 \end{bmatrix} = \begin{bmatrix} s_1 & s_2 \\ s_2^* & -s_1^* \end{bmatrix} \begin{bmatrix} h_1 \\ h_2 \end{bmatrix} + \begin{bmatrix} n_1 \\ n_2 \end{bmatrix} \quad [6.17]$$

The received signal has the following form:

$$\boldsymbol{r} = \begin{bmatrix} r_1 & r_2 \end{bmatrix}^{\mathrm{T}} = \left[\left(h_1 s_1 + h_2 s_2 + n_1 \right) \quad \left(h_1 s_2^* - h_2 s_1^* + n_2 \right) \right]^{\mathrm{T}} \quad [6.18]$$

The receiver estimates the complex gain of each channel, i.e. $\tilde{\boldsymbol{h}} = [\tilde{h}_1 \quad \tilde{h}_2]^{\mathrm{T}}$ and performs space-time decoding of the received sequence $\boldsymbol{r}$ using the estimated values of the channel gains:

$$\tilde{\boldsymbol{s}} = \begin{bmatrix} \tilde{s}_1 \\ \tilde{s}_2 \end{bmatrix} = \begin{bmatrix} \tilde{h}_1^* & -\tilde{h}_2 \\ \tilde{h}_2^* & \tilde{h}_1 \end{bmatrix} \begin{bmatrix} r_1 \\ -r_2^* \end{bmatrix} \quad [6.19]$$

$$\tilde{\boldsymbol{s}} = \begin{bmatrix} \tilde{h}_1^* & -\tilde{h}_2 \\ \tilde{h}_2^* & \tilde{h}_1 \end{bmatrix} \begin{bmatrix} h_1 s_1 + h_2 s_2 + n_1 \\ -\big(- h_1 s_2^* + h_2 s_1^* + n_2 \big)^* \end{bmatrix} \quad [6.20]$$

$$\tilde{\boldsymbol{s}} = \begin{bmatrix} \tilde{h}_1^* h_1 s_1 + \tilde{h}_1^* h_2 s_2 + \tilde{h}_1^* n_1 - \tilde{h}_2 h_1 s_2^* + \tilde{h}_2 h_2 s_1^* - \tilde{h}_2 n_2 \\ \tilde{h}_2^* h_1 s_1 + \tilde{h}_2^* h_2 s_2 + \tilde{h}_2^* n_1 + \tilde{h}_1 h_1 s_2^* - \tilde{h}_1 h_2 s_1^* - \tilde{h}_1 n_2 \end{bmatrix} \quad [6.21]$$

If the receiver exactly estimates the complex gains of the channel and that they are independent, then:

$$\begin{aligned} \tilde{\boldsymbol{s}} &= \begin{bmatrix} h_1^* h_1 s_1 + h_2 h_2 s_1^* + h_1^* n_1 - h_2 n_2 \\ h_2^* h_2 s_2 + h_1 h_1 s_2^* + h_2^* n_1 - h_1 n_2 \end{bmatrix} \\ &= \begin{bmatrix} \big(|h_1|^2 + |h_2|^2\big) s_1 + \underbrace{h_1^* n_1 - h_2 n_2}_{\breve{n}_0} \\ \big(|h_1|^2 + |h_2|^2\big) s_2 + \underbrace{h_2^* n_1 - h_1 n_2}_{\breve{n}_1} \end{bmatrix} \end{aligned} \quad [6.22]$$

where $\breve{n}_0$ and $\breve{n}_1$ represent the noise contribution at the space-time decoder output. The term $(|h_1|^2 + |h_2|^2)$ thus indicates that signals s_1 and s_2 are received in transmission diversity.

Figure 6.9 shows the MIMO system using Alamouti's code with, this time, transmitter and receiver diversity. The transmission of this same space-time signal inside a MIMO channel with two receiving antennae ($N_T = 2$, $N_R = 2$) leads to four received signals which can be exploited using a space-time processing. The complex gain matrix $\boldsymbol{H}$ of the MIMO channel is:

$$\boldsymbol{H} = \begin{bmatrix} h_{1,1} & h_{1,2} \\ h_{2,1} & h_{2,2} \end{bmatrix} \quad [6.23]$$

The receiver receives the four signals coming from the two receiving antennae at two discrete instants $k = 1$ and $k = 2$:

$$\boldsymbol{R} = \begin{bmatrix} r_{1,1} & r_{1,2} \\ r_{2,1} & r_{2,2} \end{bmatrix} \quad [6.24]$$

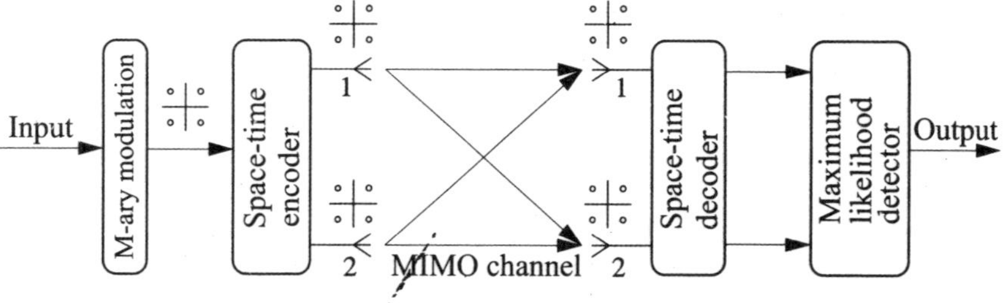

Figure 6.9. *Alamouti's code with transmitter and receiver diversity*

and estimates the complex gains of the MIMO channel: $\tilde{\boldsymbol{H}}$. Assuming that the four channels are independent and perfectly estimated, the complex symbols transmitted by the information source are decoded by:

$$\tilde{\boldsymbol{s}} = \begin{bmatrix} \left(\left|h_{1,1}\right|^2 + \left|h_{1,2}\right|^2 + \left|h_{2,1}\right|^2 + \left|h_{2,2}\right|^2\right) s_1 + \breve{n}_0 \\ \left(\left|h_{1,1}\right|^2 + \left|h_{1,2}\right|^2 + \left|h_{2,1}\right|^2 + \left|h_{2,2}\right|^2\right) s_2 + \breve{n}_1 \end{bmatrix} \qquad [6.25]$$

and the diversity gain is $(|h_{1,1}|^2 + |h_{1,2}|^2 + |h_{2,1}|^2 + |h_{2,2}|^2)$.

Alamouti's code is an example of a space-time code using spatial diversity. In effect, it makes it possible to increase the link quality – especially its robustness against signal losses – without yet increasing the total throughput, because it is a code of rate 1. Other codes using spatial multiplexing exist, which provide a throughput increase based on the Shannon matrix capacity.

6.4. Existing architectures

6.4.1. *Frequency diversity and space-time encoding: MIMO-OFDM*

The transmission of broadband information with multiple carrier signals enables the reduction and even the elimination of inter-symbol interference by breaking down a typically frequency selective broadband signal into a group of frequency non-selective narrowband sub-carriers. The modulation spread over a number of OFDM[15] orthogonal carriers constitutes one of the most interesting and widespread methods (also used for certain variants of Wi-Fi and Wi-Max systems, for DAB[16]

15 OFDM: *Orthogonal Frequency Division Multiplexing*.

16 DAB: *Digital Audio Broadcasting*.

digital radio and for DVB[17] and ISDB[18] digital television) for multiple carrier transmissions.

The advantages of OFDM modulation can be exploited in a system configuration with MIMO multiple antennae by the addition of frequency diversity to the MIMO time and space diversity. Figure 6.10 shows the basic architecture of a MIMO system with OFDM-type multiple carriers. The detection of MIMO-OFDM symbols can be done with a maximum likelihood detector using the Euclidian metric [BOL 02, PAU 03].

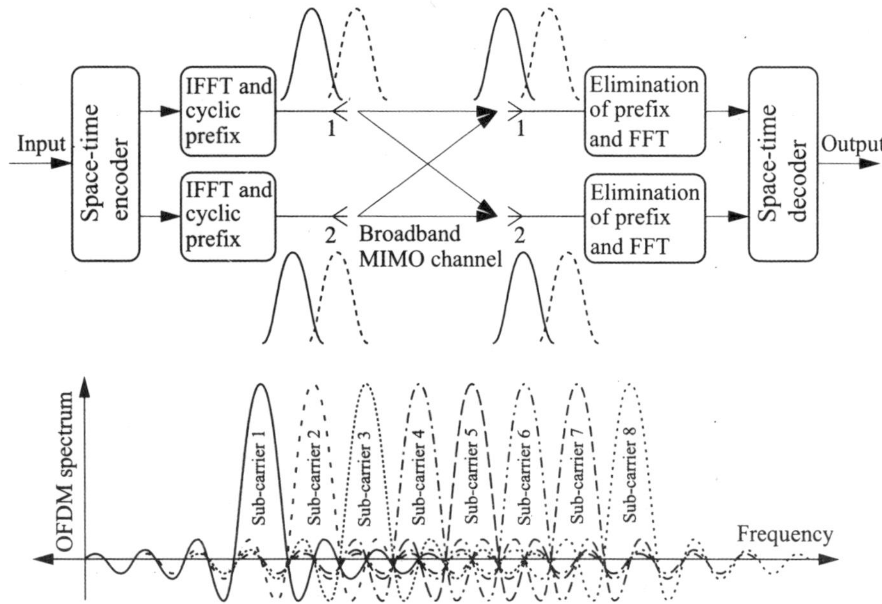

Figure 6.10. *MIMO-OFDM transmission system*

6.4.2. *Spatial multiplexing: BLAST systems*

It is possible to use the antenna arrays of a MIMO system in order to perform spatial multiplexing in order to maximize throughput. The spatial multiplexing consists of emitting different bit streams, which have little or no redundancy (i.e. no repetitions like in Alamouti's code). Upon receipt, the signals corresponding to each user are

17 DVB: *Digital Video Broadcasting*.

18 ISDB: *Integrated Services Digital Broadcasting*.

extracted by using successive nulling and cancelling methods. The BLAST[19] systems of Bell Laboratories are a known example in this respect.

The D-BLAST[20] architecture proposed by Foschini in 1996 [FOS 96] constitutes one of the first MIMO architectures in the modern meaning of the term. It is actually a form of multi-user detection form except for the fact that the different information sources (the transmitting antennae) are considered here as all belonging to a single user whose throughput we want to increase. Besides this subtlety at the interpretation level, these multi-layer architectures, such as D-BLAST, perform the same tasks as the multi-user detection algorithms, several of them being already known. D-BLAST innovates and surprises at the complexity level because it is linear with the number of transmitting and receiving antennae N_T and N_R.

The transmission matrix of the (multiplexed) D-BLAST matrix signal is as follows:

$$\boldsymbol{S} = \begin{bmatrix} s_{1,1} & s_{1,2} & s_{1,3} & \cdots & 0 & 0 & 0 \\ 0 & s_{2,1} & s_{2,2} & \cdots & 0 & 0 & 0 \\ 0 & 0 & s_{3,1} & \cdots & 0 & 0 & 0 \\ \vdots & \vdots & \vdots & \ddots & \vdots & \vdots & \vdots \\ 0 & 0 & 0 & \cdots & s_{N_T-2,N} & 0 & 0 \\ 0 & 0 & 0 & \cdots & s_{N_T-1,N-1} & s_{N_T-1,N} & 0 \\ 0 & 0 & 0 & \cdots & s_{N_T,N-2} & s_{N_T,N-1} & s_{N_T,N} \end{bmatrix} \qquad [6.26]$$

The signals coming from the N_T antennae are moved forward in time: the signals of the first diagonal, i.e. $s_{1,1}, s_{2,1}, \ldots, s_{N_T,1}$, are transmitted from antenna 1, the signals of the second diagonal $s_{1,2}, s_{2,2}, \cdots, s_{N_T,2}$, from antenna 2 and so on. The diagonal structure of the D-BLAST system makes it possible to use the time and space diversity for a multi-antenna transmission so far as the number of receiving antennae N_R corresponds or exceeds N_T. The D-BLAST system makes it possible to approach the channel capacity if the channel is quasi-static in time, non-selective and affected by additive white Gaussian noise [HAY 05].

An important limitation of the D-BLAST system like the one described is that each *diagonal layer* requires an independent (temporal) encoding, thus limiting the diagonal codes, in consequence of their realization complexity, to inefficient short correction codes. In 1998, Wolniansky, Foschini, Golden and Valenzuela [WOL 98] suggested a new architecture called V-BLAST[21] making it possible to avoid certain

19 BLAST: *Bell labs layered space-time architecture.*

20 D-BLAST: *Diagonally layered space-time architecture.*

21 V-BLAST: *Vertical BLAST.*

problems associated with diagonal layers of the D-BLAST system and to thus reach a spectral efficiency of 40 bits/s per Hz by independently encoding and by transmitting each sub-sequence via a different antenna. At the receiver, a joint multi-signal detector successively eliminates the interference (in decreasing order of the signal-noise ratios of the other sub-sequences) in order to reproduce the original data.

6.4.3. *Turbo-BLAST systems*

The iterative decoding methods used with turbo correcting codes can be applied to space-time encoding. The Turbo-BLAST architecture presented by Sellathurai and Haykin [SEL 02] is based on the V-BLAST architecture. Following the style of turbo codes, Turbo-BLAST enables a performance very close to the MIMO channel capacity with a receiver architecture whose complexity remains manageable. This architecture consists mainly of a MIMO system with multiple transmitting and receiving antennae, an interleaving space-time encoding and an iterative receiver enabling the exchange of extrinsic information (reliability values and flexible information) between a unique detector SISO[22] for the signals coming from N_R receiving antennae and SISO decoders dedicated to each user. The iterative decoding of the Turbo-BLAST system leads to better performances than those obtained with the V-BLAST architecture, but not yet surpassing those of the D-BLAST system [HAY 05].

Finally, space-time multiplexing (e.g. BLAST) can be used for the space-time encoding of a unique information source by de-multiplexing the information sequence into N_T sub-sequences in order to be transmitted by the N_T transmitting antennae of the BLAST system.

6.5. Reconfigurable MIMO systems

It is clear that a MIMO system generally involves a processing complexity more important at the two ends than a simple antenna array as described in section 6.3.3. The SUBSET selection strategy can be applied during reception if the number of antennae is big enough, but can it be applied during transmission? Not without proportionally diminishing the throughput. In order to maximize the latter, it is necessary to actually simultaneously transmit at all times using the N_T antennae, and receive using at least N_T antennae. In order to benefit from a certain spatial diversity, it would be necessary that N_R be slightly larger than N_T. This thus implies a massive replication of the RF chain, A/D, D/A converters and subsequent digital processing functions (see section 6.3.3).

22 SISO decoder (or detector): *soft-input soft-output* decoder, as opposed to binary inputs and outputs after hard decision.

However, we shall note that reception is broken down into two main phases, which are always orthogonal in time (i.e. they do not take place at the same time) in the conventional architectures: (1) the estimation phase, which determines the relevant statistical parameters; (2) the detection phase, which is based on the aforementioned parameters in order to estimate which symbols have been transmitted. This suggests the application of dynamic reconfiguration.

6.6. Case study

In this section we present two examples of applications of a MIMO antenna system that can benefit from the flexibility of reconfigurable architectures: multiple access MIMO systems with MIMO-WCDMA[23] as well as adaptive antenna architectures.

6.6.1. *MIMO-WCDMA receiver*

The multiple access technology with WCDMA is based on the dominant standard within the designated IMT-2000[24] framework of 3G cellular telephony. Consequently, this standard is network oriented (packet transmission) and has a multi-service character (e.g. multimedia). In particular, it makes it possible to manage the quality of service[25], variable throughputs based on the requirement and the coexistence of time (TDD)[26] and frequency (FDD)[27] division duplexing methods. The standard uses a frequency spacing of 5 MHz with rapid power control on the uplink and downlink. In addition, it stipulates the use of transmission diversity on the downlink with Alamouti's space-time code presented in section 6.3.6, thus opening the door for the development of more complex, higher performance MIMO and space-time systems.

By combining the MIMO multi-antenna architecture with beamforming, it is possible to significantly increase the capacity of the WCDMA system. In [BEL 04], a WCDMA system implementation project is presented[28], using a MIMO antenna array structure. Figure 6.11 shows the block diagram of the MIMO-WCDMA transmitter with two transmitting antennae. Each antenna transmits a different binary sequence with a distinct channeling code.

---

23 WCDMA: *Wideband Code Division Multiple Access*.

24 IMT-2000: *International Mobile Telephony 2000*.

25 QoS: *Quality of Service*.

26 TDD: *Time-Division Duplexing*.

27 FDD: *Frequency-Division Duplexing*.

28 The project is realized collectively at the Radio-Communication and Signal Processing Laboratories (LRTS) of Laval University and by the research group R2D2 at the Computer Science and Random Systems Research Institute (IRIS).

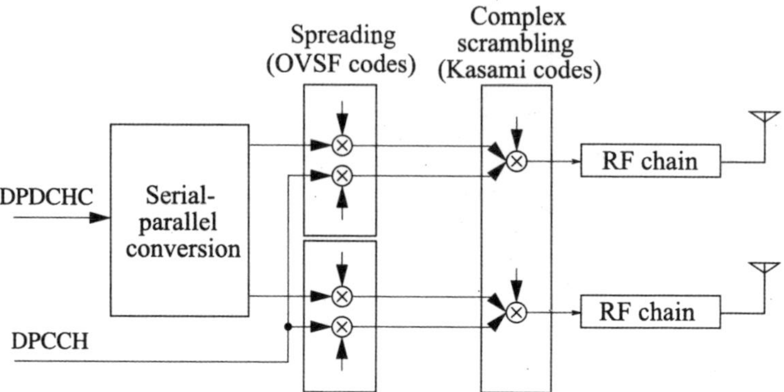

Figure 6.11. *MIMO-WCDMA transmitter*

Figure 6.12 shows the structure of the receiver with two receiving antennae. A root raised cosine filter – indicated as RRC – respecting the UMTS standard for matched filtering is used. The received sequences are over-sampled at the WCDMA chip rate. A processing unit tries to estimate the four strongest paths of the broadband signal impulse response. A *rake* detector performs the channel estimation, the WCDMA sequence despreading and the phase realignment of the different received signals. The samples at the output of the *rake* filters are then used by the MIMO decoder in order to remove the mutual interference of the signals received.

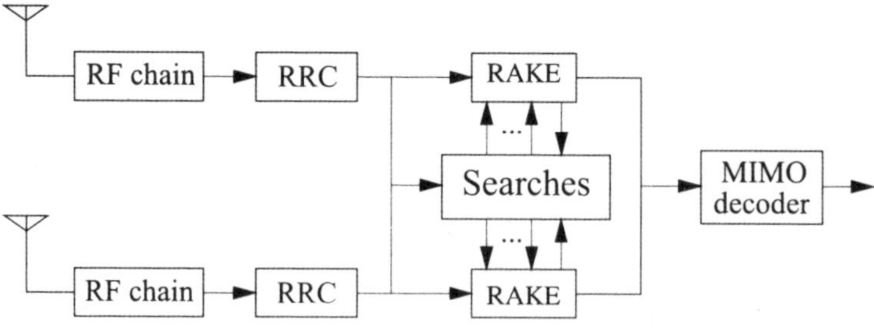

Figure 6.12. *MIMO-WCDMA receiver*

Figure 6.13 shows the internal structure of the top *rake* receiver (the other one is obviously identical). After having been despread, the digitized received signal enters a register queue for path analysis. The path search module ("searches") is common because it can use the correlations that already exist between the two channels in its function which consists of roughly isolating the most energetic paths. Each *rake* receiver has a number of branches (rake fingers) which are positioned on the most

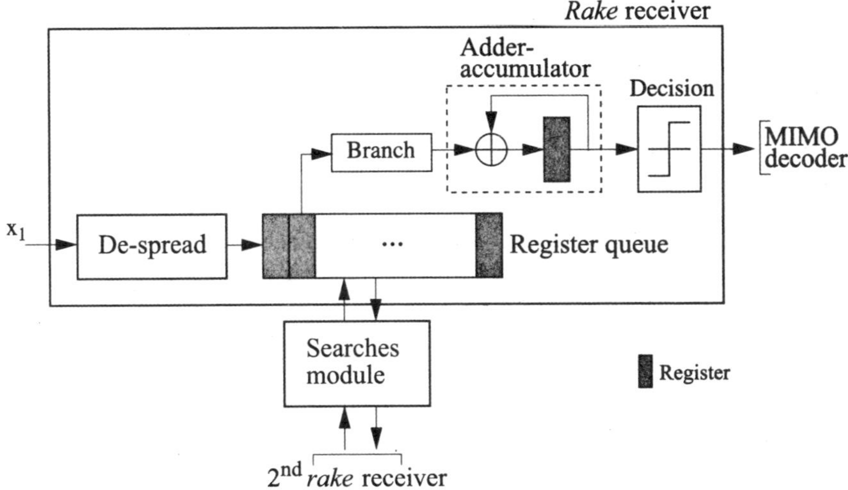

Figure 6.13. *Simplified structure of the rake receiver*

energetic paths. Each branch implements fine tracking following the movements of the path peak which was assigned to it by the research module. In the implementation strategy chosen, only two branches are in fact created (and only one is presented in the figure). However, each one of them performs the work of three branches (and follows therefore three paths) by temporal multiplexing. This is why the branch in Figure 6.13 is followed by an adder-accumulator which cumulates the contributions of the three paths.

This system has been synthesized by targeting a Virtex XCV2000E FPGA Xilinx with the help of synthesis software Synplify Pro. The RRC filter occupies 620 logic stages which corresponds to 3% of the chip logic resources. The searches module takes up about 10% of the resources (2,230 logic stages) while the branch takes up 7% (1,447 stages). The essential parts of a two-antenna receiver take up three-quarters of the chip (74% or 15,900 stages).

A hardware prototyping platform is being developed at LRTS[29] which will serve, among other things, as a support for the MIMO-WCDMA prototype system. The platform consists of three stations of four antennae, each one of them operating within the frequency bands at 2.4 GHz and 5.15 GHz with a maximal band width of 40 MHz. The processing capacity is provided by complex FPGAs (three per station) and by

29 LRTS: Laboratoire de Radiocommunication et de Traitement du Signal.

digital signal processors (four per station). The whole is coupled to a fast prototyping software suite, enabling the fast implementation of a given architecture based on predefined blocks in the Mathworks Simulink software. Since the hardware components for acquisition and processing are entirely generic, the result is a development platform which is completely integrated with the development software.

These stations will also serve as a support for the prototype of the MIMO-WCDMA architecture. The system-level design approach used, due to the constant evolution of programmable methodologies and logic, is particularly suitable for the reconfigurable architectures and software-defined radio.

6.6.2. *Receiver architectures for adaptive antenna arrays*

The simplest form of antenna processing in order to obtain diversity is the selection diversity where a single RF chain is used and connected in turn through switching to each of the antennae in order to choose the one which provides the best SNR. Although simple and not expensive, this method provides a certain diversity gain against the multipath signal losses, but it proves to be rather inefficient in the presence of interferers.

Generalized selection diversity offers a better cost/performance compromise by enabling the selection of an array subset whose antennae are, in turn, jointly combined and processed in order to obtain the desired signal. However, such a method basically needs a continuous monitoring of the SNR in all branches, which implies a certain hardware complexity, i.e. a device for power measurement inside each branch. Moreover, the processing unit must constantly read the power measurements, re-evaluate the subset optimality and, if necessary, initiate a switching sequence towards another subset.

Based on this principle, a new architecture of the reduced complexity multi-antenna receiver has been proposed in [ROY 03]. Here are the salient features:

– subset selection is performed based on its global performances and not on the SNR of the individual branches;

– selection is made based on the long term statistics of the channels (not the rapid fluctuations due to multipath propagation), in order to considerably reduce the workload associated with the continuous selection process;

– signal arrival geometry is used through a network of directional antennae. In this manner, the antennae themselves can perform spatial pre-filtering and eliminate a part of the interference. Equally, an omnidirectional antenna array coupled to a Butler matrix [BUT 66] can be used in order to create orthogonal directional beams among them.

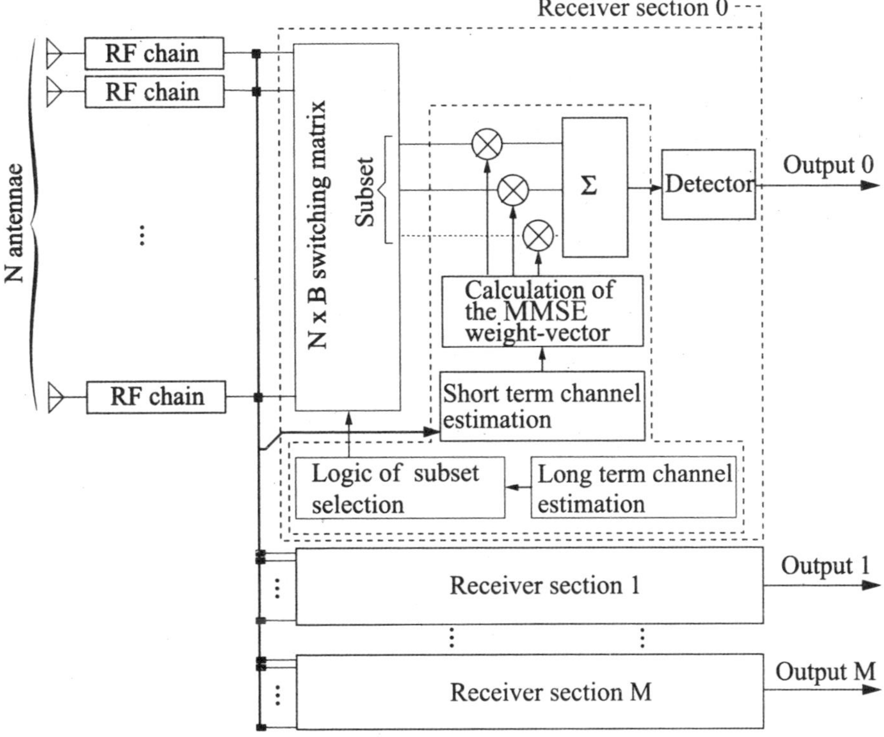

Figure 6.14. *Subset receiver section architecture designed to operate in a multi-user context*

Let us take the receiver given in Figure 6.14. In addition, we assume a fixed subset size S so that an array of N antennae should have $N_S = \binom{N}{S}$ possible subsets. The selection criterion pertains to maximizing the potential SNR as defined by averaging over a relatively long period as compared to short term fading. For a narrowband receiver operating within a flat fading environment, the selection criterion can be expressed:

$$s_o = \max_s \left\langle {\boldsymbol{c}_0^{(s)}}^{\mathrm{H}} {\boldsymbol{R}_{I+N}^{(s)}}^{-1} \boldsymbol{c}_0^{(s)} \right\rangle_t, \quad \text{where } s = 1, \ldots, N_S \qquad [6.27]$$

where $\boldsymbol{c}_0^{(s)}$ is the signature (vector channel) of the desired signal observed over subset s, $\boldsymbol{R}_{I+N}^{(s)}$ is the interference + noise covariance matrix as observed over subset s and $\langle\cdot\rangle_t$ expresses at the same time the expectation (assembly average) and the time averaging needed to obtain long term statistics.

Once the subset is chosen, an adaptive (short term) processing is performed on the chosen branches. This operation can for example aim to minimize the MSE, which corresponds to the maximization of signal-to-interference-plus-noise ratio (SINR).

Thus, the weight vector of an ideal combiner should be:

$$\boldsymbol{w}^{(s_o)} = \boldsymbol{R}_{I+N}{}^{-1}\boldsymbol{c}_0^{(s_o)} \qquad [6.28]$$

The simulation of such a system has been performed considering an array of size $N = 6$, a subset size $S = 2$, five carriers per sector, 120° sectors (such as those currently used in cell phone networks) and a cell radius of 1 km. In addition, the system allows a maximum of two users to coexist on the same carrier at the same time, based on the spatial discrimination capacity of the array. Moreover, a total occupancy (ten active users per sector) is assumed.

The simulation performed is characterized by a two-level Monte-Carlo structure. At the first level (slow variations), the positions of the users are randomly generated. The angular dispersion at arrival is set at 22° and it is assumed that the energy is uniformly distributed inside this cone. At the second level (short term), a Monte-Carlo simulation is performed on the Rayleigh multipath fading in order to establish the average performance of the system for a given geographical distribution of users.

Moreover, we use a carrier distribution strategy which considers the network capacity to separate user pairs based on spatial discrimination. The strategy is simple: the arrival directions are sorted in ascending order. In order to maximize the angular distance of the users on the same carrier, carrier 1 is assigned to the users having the first and fifth arrival directions, carrier 2 to the users having the second and sixth arrival directions, etc.

The results are shown in Figure 6.15 and they are compared to conventional antenna arrays, i.e. fully adaptive. We supposed that the conventional arrays have a sufficient inter-element spacing in order to ensure fading decorrelation. The subset selection array is constrained to a $\lambda/2$ spacing because of the use of Butler matrices by way of spatial pre-filtering; this means that it has a disadvantage with respect to fading as compared to the conventional arrays used. Despite this, the system enables a 8-9 dB gain compared to the two antenna conventional receiver and a significant gain compared to the three antenna conventional receiver when the SNR is weak. It should be noted that the complexity of the receiver proposed with $N = 6$ and $S = 2$ only slightly exceeds the complexity of the two antenna conventional receiver.

6.7. Conclusion

The advantages offered by antenna arrays led the international standard setting bodies to propose MIMO systems and space-time codes in some of their future standards. Hence, the future standard IEEE 802.11n anticipates the use of multi-carrier space-time encoding technologies with a MIMO-OFDM antenna array. However, we should not underestimate the difficulties represented by the adjustment of the space-time processing reconfigurable MIMO transceivers. Among other things, the MIMO systems need an environment which is rich in scatterers in order to reach their full

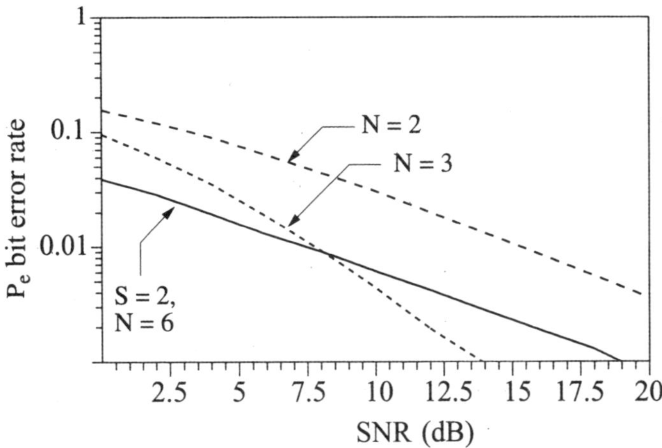

Figure 6.15. *Relative performance in terms of bit error rate of conventional antenna array receivers minimizing the MSE with two and three antennae (dotted) and of the proposed receiver with six antennae and a subset size of 2 (bold)*

potential. The imperfect estimate of MIMO channel conditions, the correlation existing between the multipath propagation trajectories, the interference between symbols and the rapid variations of signals to be processed at the receiver are many factors affecting the quality of received signals [MIE 04]. The practical limitations of reconfigurable systems such as the imperfections of A/D converters [REE 02], the number of components, the complexity of calculations and the constraints (partition and delays) on dynamic reconfiguration equally affect the performances of these systems. Nevertheless – given the capacity pressures applied to the wireless systems, the coexistence of multiple standards within the same bands and the speed of technological evolution – it is clear that antenna arrays, MIMO systems and especially their reconfigurable versions are inevitable steps in the future evolution of wireless communications. Consequently, and taking into account the challenge they represent, the study of these systems will constitute one of the most relevant and most fascinating research topics for the foreseeable future.

6.8. Bibliography

[ALA 98] ALAMOUTI S., "Space block coding: a simple transmitter diversity technique for wireless communications", *IEEE Journal on Selected Areas in Communications*, Vol. 16, p. 1451–1458, October 1998.

[BEC 63] BECKMANN P., SPIZZICHINO A., *The Scattering of Electromagnetic Waves from Rough Surfaces*, MacMillan, New York, 1963.

[BEL 04] BÉLANGER L.N., ROY S., SAÏDI T., SENTIEYS O., "Prototyping a MIMO W-CDMA system using a system-level approach", in *Global Signal Processing Conference and Expo (GSPx)*, Santa Clara, September 2004.

[BOL 02] BÖLCSKEI H., GESBERT D., PAULRAJ A., "On the capacity of OFDM based spatial multiplexing systems", *IEEE Transactions on Communications*, Vol. 50, No. 2, p. 225–234, February 2002.

[BUT 66] BUTLER J.L., *Digital, Matrix and Intermediate Frequency Scanning*, Vol. 3, p. 217–288, Academic Press, New York, 1966.

[ETS] ETSI: European telecommunication standards institute, 1™ETSI website, http://www.etsi.org.

[FOS 96] FOSCHINI G.J., "Layered space-time architecture for wireless communications in a fading environment when using multielement antennas", *Bell Labs Technical Journal*, p. 41–59, August 1996.

[HAY 05] HAYKIN S., MOHER M., *Modern Wireless Communications*, Pearson Prentice-Hall, Upper Saddle River, 2005.

[HON 97] HONCHARENKO W., KRUYS J.P., LEE D.Y., SHAH N.J., "Broadband wireless access", *IEEE Communications Magazine*, Vol. 35, No. 1, January 1997.

[IEE] IEEE: Institute of electrical and electronics engineers, 1™IEEE website, http://www.ieee.org.

[JAK 74] JAKES W.C., *Microwave Mobile Communications*, IEEE Press, Piscataway, 1974.

[MIE 04] MIETZNER J., HOEHER P.A., "Boosting the performance of wireless communication systems: theory and practice of multiple-antenna techniques", *IEEE Communications Magazine*, Vol. 42, No. 10, p. 40–47, October 2004.

[PAU 03] PAULRAJ A., NABAR R., GORE D., *Introduction to Space-Time Wireless Communications*, Cambridge University Press, Cambridge, 2003.

[REE 02] REED J.H., *Software Radio: A Modern Approach to Radio Engineering*, Prentice-Hall Communications Engineering and Emerging Technologies, Pearson Prentice-Hall, Upper Saddle River, 2002.

[ROY 03] ROY S., "Reduced complexity array receiver with subarray selection", in *IEEE Pacific Rim Conference on Communications, Computers, and Signal Processing (PACRIM™2003)*, Victoria, Canada, p. 748–751, May 2003.

[ROY 05] ROY S., NÉRON J.S., "The impact of estimation error on the performance of predetection selection diversity combining", *Wireless Personal Communications*, 2005.

[SEL 02] SELLATHURAI M., HAYKIN S., "Turbo-BLAST for wireless communication: theory and practice", *IEEE Transactions on Signal Processing*, Vol. 50, p. 2538–2546, 2002.

[VER 03] VERKEST D., "Machine chameleon: a sneak peek inside the handheld of the future", *IEEE Spectrum*, Vol. 40, No. 12, p. 41–46, December 2003.

[WAN 01] WANG Q., LECOURS M., VERGNOLLE C., "Criteria for wide-band radial switch design", *IEEE Transactions on Microwave Theory and Techniques*, Vol. 49, No. 1, p. 128–132, January 2001.

[WIN 92] WINTERS J.H., SALZ J., GITLIN R.D., "The capacity of wireless communication systems can be substantially increased by the use of antenna diversity", at *International Conference on Universal Personal Communications (ICUPC™1992)*, p. 2.01.1–2.01.5, 1992.

[WOL 98] WOLNIANSKY P.W., FOSCHINI G.J., GOLDEN G.D., VALENZUELA R.A., "V-BLAST: An architecture for realizing very high data rates over the rich scattering wireless channel", in *Proceedings of ISSSE (Pisa, Italy)*, September 1998.

Chapter 7

Analog-to-Digital Conversion for Software Radio

7.1. Introduction

The aim in software radio is to have a transceiver reconfigurable according to the available standards and services. The ideal software radio, as presented by Mitola [MIT 95], will be available only in the very remote future (20 years or more). For the next 10 years the evolution will take place around the software radio known as restricted (*software defined radio*: SDR). The main elements constituting a transceiver of this type are the intelligent antenna, programmable RF modules, broadband and high resolution analog-to-digital and digital-analog converters, digital signal processing. In this chapter, we are interested in the reception part and more particularly in the function of analog-to-digital conversion (ADC).

The receiver's flexibility is improved by bringing the converter closer to the antenna. This evolution imposes an increase in the performances of the ADC in terms of resolution and speed. In section 7.2, we will show the current state of digitalization possibilities of the analog signal in the standard technologies. Section 7.3 will place the ADC in the various possible architectures of the current receivers. Section 7.4 will discuss various types of ADC usable for our application. Lastly, the perspectives of the ADC evolution in the next 10 years will finish this chapter (section 7.5).

Chapter written by Patrick LOUMEAU, Lírida NAVINER and Jean-François NAVINER.

7.2. Current ADC performances

For radio communication applications, the essential parameter is the signal-to-noise ratio plus distortion at the converter's output. It makes it possible to define the resolution *n* of the converter in number of effective bits (ENOB: *effective number of bits*) by the relation:

$$(\text{SNDR})_{\text{dB}} = 6.02\text{n} + 1.76 \quad [7.1]$$

The output performance of ADCs usable in the application given in number of samples per second is between 10^7 samples/s and some 10^9 sample/s. At these speeds, the limitation is due to the clock phase jitter [WAL 99].

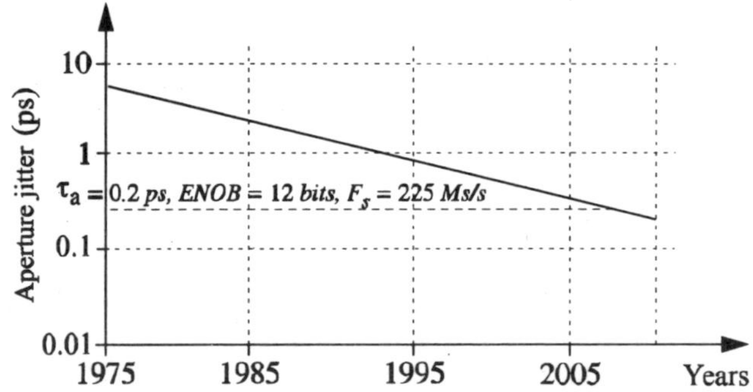

Figure 7.1. *Evolution of the aperture jitter in time*

If we suppose that the aperture jitter τ_a creates an error on the sampled signal and that this error is in the worst case equal to a half-quantum, $\left(\frac{1}{2^{n+1}}\right)$, we obtain the relation:

$$2^n = \frac{1}{\sqrt{3}\pi F_{in} \tau_a} \quad [7.2]$$

where F_{in} is the input signal frequency. The current typical value of τ_a in standard technology is about 0.5 ps, as indicated in Figure 7.1. Based on this value and on equation [7.2], we infer that the maximum resolution for 10 Mega samples/s (Ms/s) is of 15 bits and then it decreases with a slope of – 1 bit/octave and reaches 7 bits with 2.5 G/s. This gives the performance limit of the ADC in the current standard technologies. The foreseeable evolution in time of the phase jitter (see Figure 7.1) does not give significant improvement in the near future. The possible choice of the architectures of receivers will give us an indication on the expected and possible ADC performances.

7.3. Architecture of receivers

The classification of architectures of the receivers can be done according to the ADC position in the chain, as shown in Figure 7.2 [ELW 01].

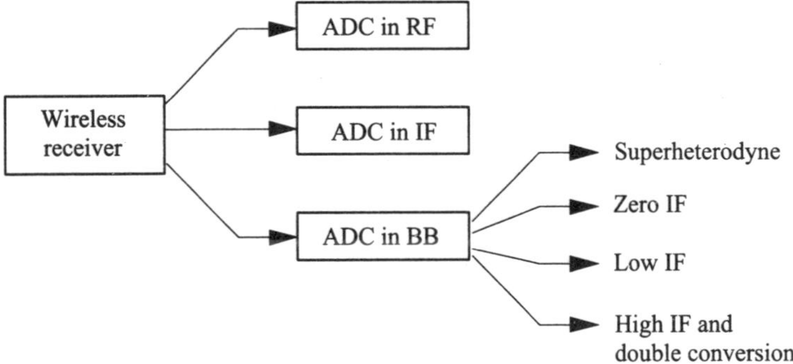

Figure 7.2. *Classification of architectures*

Placing the ADC in RF right after the antenna, as stipulated in the ideal software radio, imposes a resolution directly derived from the dynamics of the RF signal around 100 dB, that is to say an resolution equivalent to 18 bits. Moreover, the signal must be sampled at the RF frequency, that is to say potentially at several GHz. The preceding section showed the performances of current standard technologies, which are very far from these values. Indications will be given in the last section in order to obtain such specifications.

The second possibility is to place the ADC in intermediate frequency (IF) preceded by the weak noise amplifier and the mixer. In comparison to the preceding solution, the specifications are toned down and become conceivable with standard technologies at the price of a strong consumption. This places this solution at the level of base stations. This point will be developed in the next section.

The third category of receivers digitizes the signal in baseband (BB). The traditional superheterodyne receiver presents entirely analog RF and IF stages. The digital processing is carried out only in baseband. This receiver presents a good performance in selectivity, in separation between the desired channel and the other channels and in sensitivity: the minimal signal which we must be able to recover while guaranteeing the signal-to-noise ratio at the receiver's output. The major disadvantage is the obligation to use non-integrative and non-adjustable frequency filters, which are primarily surface wave filters, in RF and IF. This makes its use difficult within the framework of multi-standard applications.

Zero IF and low IF receivers enable solutions that are more integrated and more accessible to the processing of several standards. The attenuation filters of the frequency image and the intermediate frequency filters are removed. Based on the amplifier's low noise, the signal processing could be realized in the form of an integrated circuit. It is the same for the solution with broad IF and double conversion [RUD 97].

Zero IF and low IF solutions are from now on the most widely used in the mobile terminals. The selection of the channel and final amplification are in baseband. This point will be developed later with a detailed presentation of the two most widely used ADC architectures: ADC pipeline and sigma delta.

7.3.1. *Sampling in intermediate frequency*

The general architecture of a receiver with sampling in intermediate frequency is presented in Figure 7.3.

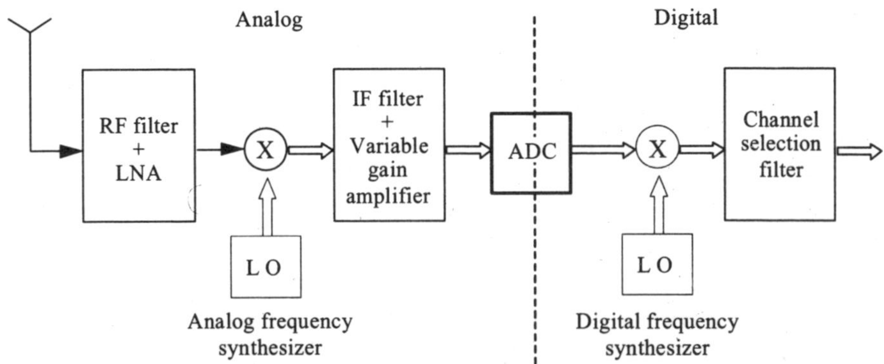

Figure 7.3. *Receiver with sampling in intermediate frequency*

The received frequency band is selected by the RF filter. This signal is then amplified by a low noise amplifier (LNA) and is relocated in intermediate frequency (IF) lower than the radio frequency. With the help of the mixer, the local oscillator (LO) will overlap, on the intermediate frequency, the signals above and below its value. This creates the problem of the frequency image. To avoid it, the signal is processed in complex form by creating the two paths, one in-phase (I) and one quadrature (Q) channels, as represented in Figure 7.3 by double arrows.

At the ADC's output, the translation towards the baseband, the filtering channel and the output of the digital data are done in the digital field. This type of receiver works with a high clock frequency. This has several advantages. First of all, for a desired signal with a fixed band, a higher sampling rate increases the transition band (instantaneous frequency deviation between the stopband and the bandwidth) and decreases the complexity of the analog filter, which is placed before the converter. Another advantage is the effect of oversampling which we can identify with a gain:

$$G_{OSR} = 10\log\left(\frac{F_s/2}{B}\right) \quad [7.3]$$

with F_S as the sampling frequency, B the useful band of the signal and we suppose that the noise outside the desired band is removed by the digital filtering which follows. This gain can further be increased by using a band-pass sigma delta modulator. This solution is presented in the section on ADC architectures.

The wanted signal can be placed in various Nyquist zones, as indicated in Figure 7.4. The only constraint is the speed of the clock which must be at least twice the width of the band of the desired signal and the ADC must admit signals at the maximum frequency of the wanted signal.

The example below will give some orders of magnitude

Here are the data for an ADC:

– resolution 14 bits;

– speed 100 Ms/s;

– input signal sampling up to 200 MHz.

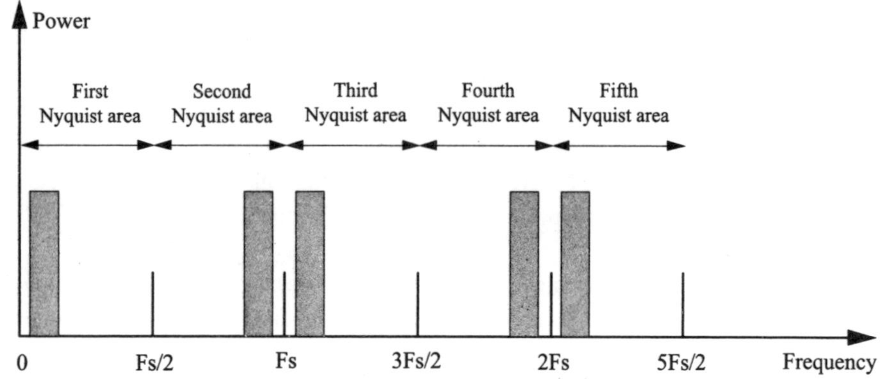

Figure 7.4. *Example of wanted signal positioning with respect to the sampling frequency*

If we suppose that the wanted signal occupies a 5 MHz band, we can imagine several possibilities:

– the intermediate frequency is in the second Nyquist zone and the carrier has a value of 90 MHz. In this case, at the converter's output we find the signal centered around 10 MHz;

– the intermediate frequency is in the fourth Nyquist zone and the carrier has a value of 190 MHz. In this case, we find the output signal at the same place as previously.

This makes it possible to illustrate this sampling possibility at an intermediate frequency. The price to be paid is consumption. For example, the Analog Devices AD6645 converter, with specifications similar to the preceding example, dissipates 1.5 W. This consumption does not make it possible to use this solution for mobiles and we will prefer a type of receiver architecture in direct conversion in order to convert the signal directly in baseband and to obtain much lower consumptions.

7.3.2. *Zero IF or low IF receiver*

The architecture of this type of receiver is presented in Figure 7.5. The two diagrams presented, (a) and (b), depend on the type of ADC used. The channel filtering is in the analog or digital part. This solution is very favorable to a monolithic integration and to multi-standard applications.

In the zero IF structure, which is also called direct homodyne conversion, the RF spectrum is directly brought back in baseband by eliminating the image rejection and intermediate frequency filters. However, there remains the well-known problem of the continuous and noise component in 1/f. The effect can be reduced by a negative feedback in the analog part or elimination in the digital part [RAZ 97].

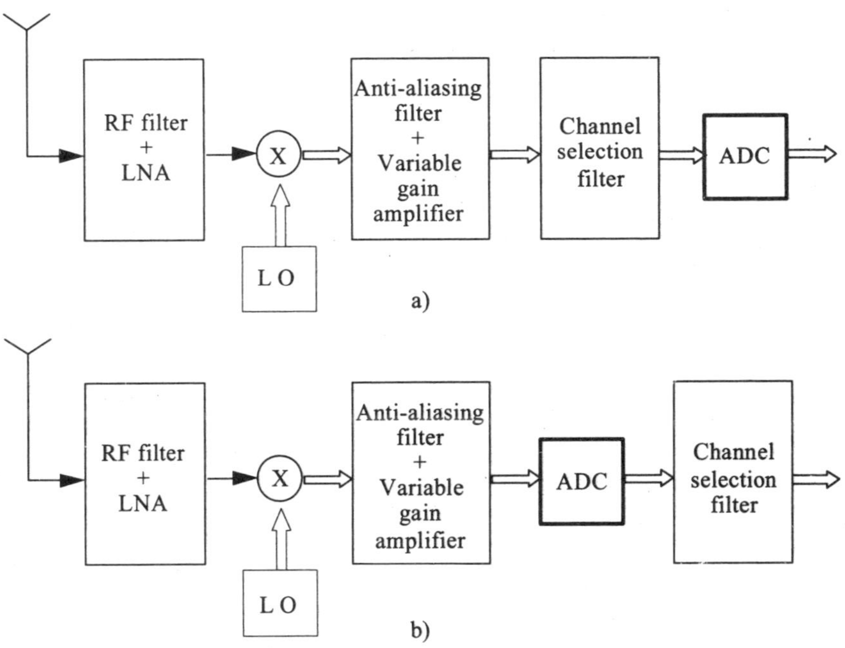

Figure 7.5. *Architecture of IF zero or low IF receiver*

For the low IF receiver, the RF signal is relocated in low frequency, but not to zero. This avoids the preceding problem, but not that of the image frequency. Mixer techniques with image rejection are used [RAZ 03].

Following this presentation of various architectures of usable receivers, the following section is dedicated to the operation of analog-to-digital conversion.

7.4. ADC architectures

Currently, in RF reception, two types of converters are used: pipeline type converters and sigma delta converters. Both fit the specifications desired in terms of speed and resolution. The performance evolution of this function will be presented at the end of the chapter.

In this section, we will present the principles of these two types of converters, their fundamental architectures and also their limitations.

7.4.1. *Analog-to-digital pipeline converter*

7.4.1.1. *Principle of pipeline conversion*

Pipeline converters are particularly interesting when a broad bandwidth and an average to high resolution are necessary. The principle consists of carrying out the conversion in successive stages (Figure 7.6).

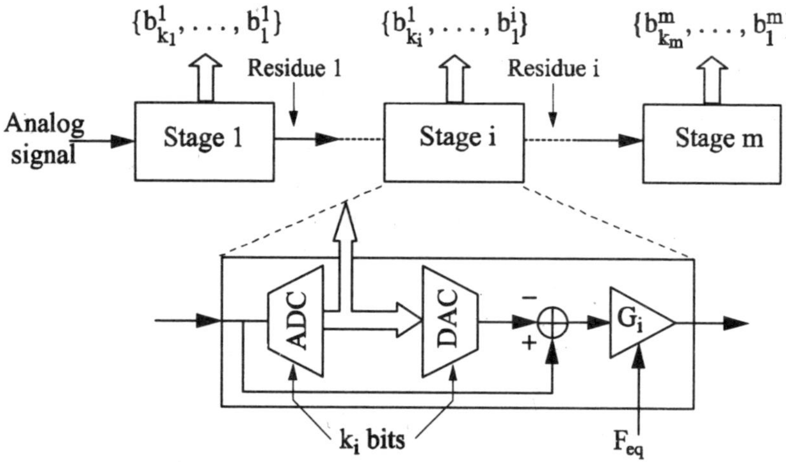

Figure 7.6. *Principle of the pipeline converter*

At the beginning, a rough conversion provides the most significant bits (MSB). The residue of this operation is calculated. The residue is the difference between the analog signal at the input of the first stage and the analog value obtained from a digital-analog conversion of MSB bits. This residue thus constitutes the input of a second stage, similar to the first, which also carries out a partial conversion, hence providing the following bits of weaker weight. In the same way, a new residue can be calculated and transmitted to the following stage, until obtaining the total desired resolution. Each stage is separated from the following one by a resampling. This enables the concomitant data processing in each stage which enables a high sampling rate.

In the ideal case (when all the components are perfect), the composition of the digital code is obtained by simple concatenation of the partial codes resulting from the various stages. With the help of the notations in Figure 7.6, the code has the following form:

$$\{b_{k_1}^1,\cdots,b_1^1,\cdots,b_{k_i}^i,\cdots,b_1^i,\cdots,b_{k_m}^m,\cdots,b_1^m\} \qquad [7.4]$$

Each stage is thus composed of an ADC, a DAC, amplification, difference and sampling functions. Amplification is used to bring back the dynamics of the residue to the full scale. In practice, the DAC, amplification, difference and sampling functions are generally associated within the same circuit called multiplying DAC (MDAC). In the ideal case, the gain applied corresponds to the number of possible codes of the stage. The ADC is carried out with a flash type differential architecture on resistive or capacitive scale. This is very traditional and partially covered by [LEW 87]. For low resolutions by stage and with the help of the introduction of architectural improvements developed later on, the comparators can be very simplified and with low consumption.

7.4.1.2. *Errors of pipeline converters*

In practice, since the used components are imperfect and sources of noise or sensitive to noise, the performance of the converters are limited. The main sources of error are: finite gain, response time and slew-rate for the operational amplifier, mismatch, limited band and thermal noise. These causes induce, among others, gain errors, continuous component errors and non-linearities on the partial ADC and DAC [LEW 87].

The continuous component errors and gain errors as well as ADC non-linearities can be largely reduced by the introduction of redundancies and digital correction. DAC non-linearities on the other hand require calibration (generally self-calibration) and an adequate dimensioning of the conversion chain.

7.4.1.3. *Redundancy and digital correction*

The redundancy introduction consists of making the conversion fields of two neighboring stages partially common: stage i provides k_i effective bits plus one or several bits of weak weights redundant with respect to the bits of strong weights calculated at stage i + 1. For that purpose, the gain of the MDAC of a stage from now on is made lower than the total number of its possible codes. The redundancy introduction makes sense only between two consecutive stages. No modification is thus carried out on the last stage.

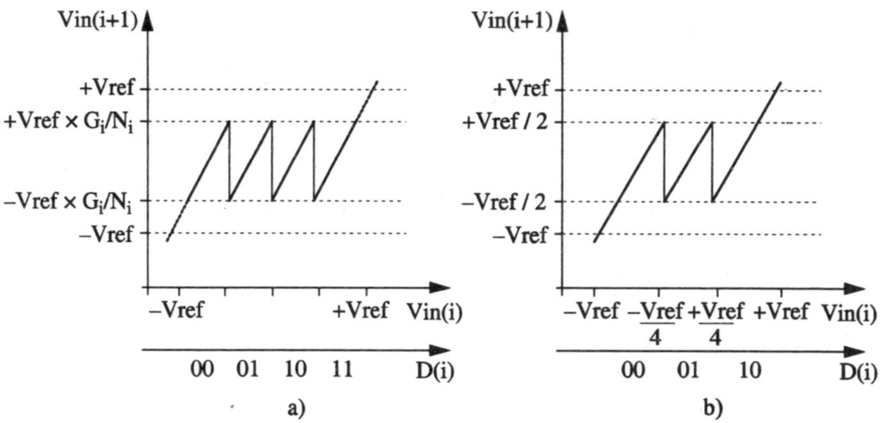

Figure 7.7. *Transfer functions of a pipeline stage: (a) 1 effective bit and 1 redundancy bit (b) stage with 3 possible codes (RSD)*

Figure 7.7(a) represents the transfer function of a stage producing 1 effective bit and 1 redundancy bit. The gain provided is 2 here, whereas the number of codes is 4. The dynamics of the output signal Vin (i + 1) is thus half of the full scale in the absence of error. In the presence of a defect (shift of the comparator threshold, for example), the error produced on the transfer function is recoverable on the following stage as long as the characteristic does not exceed the limits of the full scale.

Figure 7.7(b) represents an improvement suggested in [GIN 92, LEW 92]. A comparator, and therefore also a code, are removed, but the resolution remains unchanged. This technique is known under the name of *redundant signed digit* (RSD).

The digital correction consists of, in the majority of the cases, adding the partial codes of the stages after the shift according to the weight of the codes. An example is presented in Figure 7.8 where a two-stage converter is considered. The first stage operates a 1.5 bit conversion and the second stages a 3 bit conversion. The gain between the two stages is equal to 2. We consider that the first stage provides an erroneous decision by choosing code 2 (i.e. 10) instead of code 1 (code 01) for the Vin value at the input. The amplified residue of the first stage corresponds to code 1 of the second stage (i.e. 001). The code produced by the first stage is shifted with 3 bits and then added to the code of the second stage, thus providing the total correct code 1001.

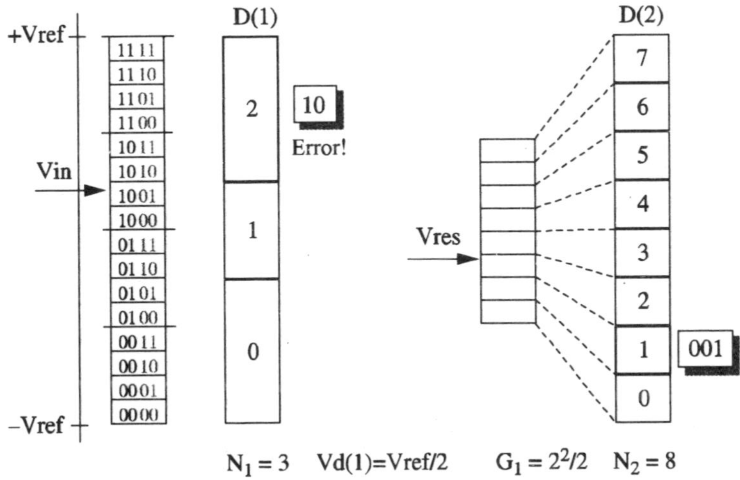

Figure 7.8. *Example of RSD pipeline converter*

By introducing redundancy and digital correction, big errors of the partial ADC are corrected. However, this technique does not have any effect on the partial DAC non-linearities. In order to correct the DAC non-linearity, apart from a suitable dimensioning of the stages, self-calibrating techniques are used, at least for the strong weight stages [GOE 94]. Indeed, if we note by e_i the error and by G_i the gain introduced in stage i, the total error brought back to the input can be written:

$$e_{in} = e_1 + \sum_{i=1}^{m-1} \frac{e_{i+1}}{\prod_{j=1}^{i} G_j} \qquad [7.5]$$

This equation shows that the first stages bring the strongest contribution to the error. Many calibration techniques are proposed, both analog and digital. The process always consists of measuring the non-linearity errors, to store the information and then, finally, to correct them [LEE 92]. A total or partial calibration is sometimes applied during the normal conversion operation but leads then to an increased consumption [SOE 95, MIN 00]. Additional circuits are necessary for calibration: analog and digital multiplexers, conversion charts, RAM, adders, etc.

7.4.2. *Analog-to-digital converter with sigma delta modulation*

7.4.2.1. *Introduction*

The sigma delta modulation converters use two fundamental techniques: oversampling and sigma delta modulation which, when they are combined, provide an effective quantization noise deletion in bandwidth at the expense of a delay on the signal. A compromise can be obtained between the resolution of the quantifier, the sampling rate and the modulation order, in order to obtain a signal-to-noise distortion ratio (SNDR). Sigma delta modulation requires a digital post-processing in order to filter the noise outside the band and to bring back the signal to the Nyquist frequency. A major interest of this technique is that the complexity of the analog part is reduced (at the expense of digital processing) and the characteristics of many analog components are toned down [LOU 02]. The sigma delta converters (SDADC) are competing for a wide range of applications in the field of telecommunications. We present in this section the principles of sigma delta modulation, the fundamental architectures and some limitations.

7.4.2.2. *Sigma delta modulation and oversampling*

Figure 7.9 provides a general representation of a sigma delta converter.

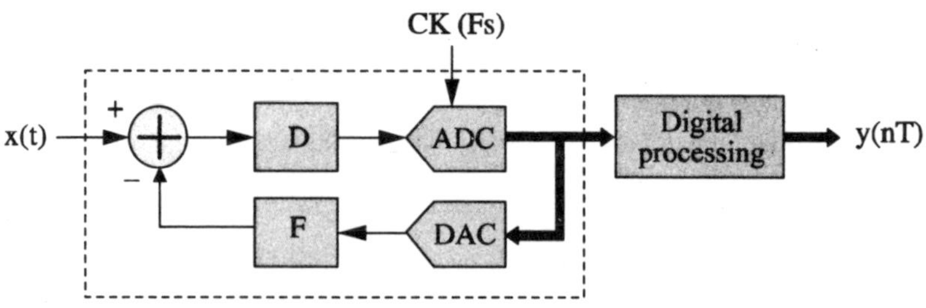

Figure 7.9. *Sigma delta converter*

We suppose here that the D block is an N order integrator filter, that the F block does not modify the signal which crosses it, that the ADC quantifier is uniform and that it can be modeled by the addition of a white noise signal. In this case, the signal to quantization noise ratio (SQNR) is described by the equations:

$$SQNR = SQNR_{quantizer} + \Delta SQNR \qquad [7.6]$$

$$\Delta SQNR = 10 \cdot \log(2 \cdot N + 1) - N \cdot 20 \cdot \log \pi + L \cdot N \cdot 20 \cdot \log 2 \qquad [7.7]$$

$$SQNR_{quantizer} = \left| \begin{array}{l} 10 \cdot \log \frac{3}{2} + 20 \cdot \log \frac{S_{max}}{ref} \\ +20 \cdot \log\left(2^n - 1\right) + L \cdot 10 \cdot \log 2 \end{array} \right. \qquad [7.8]$$

where $SQNR_{quantizer}$ represents the signal to quantization noise ratio of the quantizer included in the loop and $\Delta SQNR$ represents the contribution of the sigma delta modulation. N is the order of the modulator, n is the resolution of the quantizer, ref is the reference magnitude, S_{max} is the maximum amplitude of the signal at the input and $L = \log_2 OSR$ where OSR is the oversampling rate.

The sigma delta modulation causes very strong reduction of the quantization noise in the bandwidth and rejects it around the half frequency of the sampling frequency. This characteristic leads to, for the same structure of SDADC, a compromise between the bandwidth of the signal and the resolution obtained, unlike other conversion techniques.

7.4.2.3. *Limitations*

Instability is an important problem which is likely to limit the SDADC performance. Due to the assumptions considered for calculation, equations [7.6], [7.7] and [7.8] indicate that the signal to quantization noise ratio increases as much as the amplitude of the input signal increases. In reality, for significant amplitudes (lower than S_{max}), the SQNR falls very quickly. This causes a limitation of the acceptable dynamics at the converter's input. The single loop sigma delta modulators are particularly sensitive to instability, and this all the more so since the importance of the filter.

The sigma delta modulators are more tolerant with respect to the usual imperfections of the analog components, such as operational amplifiers, switches, capacitors.

7.4.2.4. *Architectures*

Due to instability problems, single loop structures do not exceed in general a 2 or 3 order. For higher orders, cascaded structures are preferred. Each stage is thus composed of a modulator sigma delta of order 1 or 2. Like a pipeline structure, the second stage uses the residue of the first one, thus producing a new residue which will be processed by the third stage and so on. The stages are structured in order to obtain a total transfer function which is equivalent to that of a single loop structure. A cascaded structure requires a digital recombination of the outputs of each stage before filtering and decimation.

As shown in equations [7.6], [7.7] and [7.8], a significant SQNR can be obtained even when a quantization on only one bit is used. This option is often retained in practice because the quantifier is then reduced to only one comparator and the DAC in the feedback loop is reduced to two values, and is therefore intrinsically linear. However, when a significant resolution and bandwidth are necessary, multibit quantization becomes interesting. A flash type quantizer is then used. The difficulty lies in the DAC non-linearities, which are directly added to the signal and thus restored at output. Many methods are developed in order to improve the linearity of the DACs in this case.

Other fundamental architectural options are offered to the designer: the signal sampling can be carried out upstream of the loop or right at the quantization time. In this second case, we speak about continuous-time sigma delta modulation. Only the quantizer and the DAC operate in discrete-time. A continuous-time modulator does not require an upstream anti-aliasing filter, is less sensitive to sampling noise and can work at higher frequencies. This structure is, on the other hand, particularly sensitive to the uncertainties on the sampling moment, source of non-linearities [CHE 00].

As long as the sigma delta modulation is based on the use of filters, it is possible to obtain a passband modulation instead of a low-pass one by replacing the integrating filter by a resonator. This alternative is interesting for an intermediate frequency conversion and a demodulation in digital quadrature (Figure 7.10).

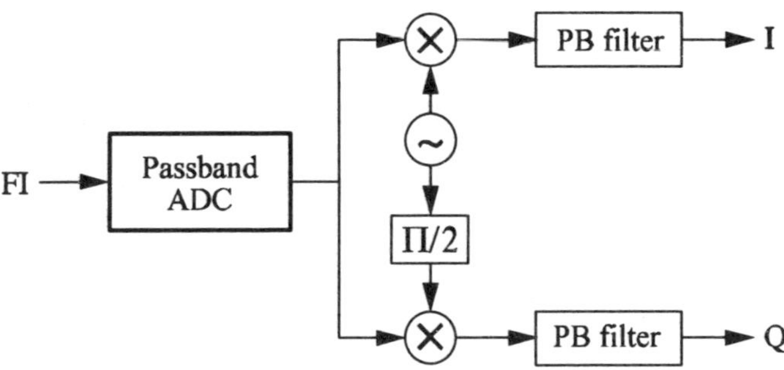

Figure 7.10. *Intermediate frequency passband conversion*

7.4.3. *Analog-to-digital converters and reconfigurability*

The reconfiguration possibilities of ADCs can influence the choices of front-end architectures. In this section we will study how pipeline and sigma delta ADC architectures can be made reconfigurable.

A pipeline converter consists of a succession of stages, each one of them carrying out a partial conversion. The number of bits of a stage is normally calculated according to bandwidth resolution and consumption specifications. A simple possibility of reconfiguration in a pipeline chain is to restrict its use to the strong weight stages when a resolution which is lower than the nominal resolution is sufficient. This involves a reduction of the consumption. When two communication standards must use the same pipeline converter, this must be dimensioned according to the broadest bandwidth and the highest resolution. There is not a possible compromise between bandwidth and resolution. The internal reconfiguration at each stage is possible but more complex to carry out.

By the spectral separation of the quantization noise (high-pass modulation) and signal, a sigma delta ADC naturally offers a compromise between bandwidth and resolution. This proves to be particularly interesting when a standard requires a broad spectral band but a moderate resolution, whereas another requires a narrower band but a high resolution. It is, for example, the case for a compatible reception chain, GSM and UMTS. This intrinsic compromise is an advantage for an ADC sigma delta architecture with respect to a pipeline architecture, since it is not necessary to consider the worst cases of resolution and bandwidth in order to size up the converter.

7.4.4. *Digital front-end: filtering for ΣΔ conversion and channel selection*

In the modern communication systems, a great number of functions are carried out with the help digital devices (channel equalization, encoding/decoding, modulation/demodulation and channel selection). In the case of broadband software radio, the channel digital selection in baseband enables the necessary adaptation in order to adapt the various bandwidths, sampling rates and CNR (*carrier over noise ratio*) constraints of the various standards of communication [MIT 95]. The digital filters play a very important part here. They must enable the selection of the information channel in the presence of interference channels and quantization noise due to the digitalization process, which is often based on the ΣΔ modulation (Figure 7.11).

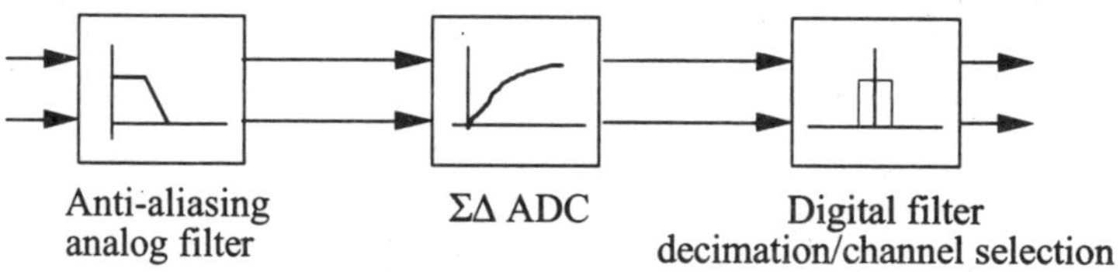

Figure 7.11. *Receiver based on sigma delta conversion: the digital filter carries out the quantization noise suppression and selects the desired channel*

An important parameter in the design of the ADC (analog-to-digital converter) is the supposed performance which is likely to come from decimation (interpolation). To this performance contribute the type, the order and the architecture of the filter used. Several factors intervene in the choice of the filter:

– factors related to the feasibility (number of coefficients, complexity, cost, consumption);

– factors related to the application constraints (minimal quality of filtering, requirement or not of linear phase).

The order of a filter (and, consequently, its complexity) depends directly on the oscillation in attenuated passband and stopband (δp and δs) and grows inversely proportional to the transition band/work frequency ratio (Δf/fs) [CRO 83]. The weak Δf/fs ratio of decimator filters for converters ΣΔ due to oversampling leads to a significant implementation cost. In order to overcome the problem of the computation power, the decimator filter can be structured in a multistage form. In this case, where several filters are cascaded, the signal is undersampled after each stage. Only the first filter works at the initial sampling rate *fsa*. As it must not carry out all the decimation, its specifications are less strict. The stricter constraints in attenuation and transition bandwidth are left to the subsequent filters, working at lower frequencies (Figure 7.12).

The comb filter is an interesting solution for the decimation process because it can be implemented without any multiplier [CHU 84]. Unfortunately, this filter has two disadvantages: it does not bring enough attenuation in rejected band and it generates a distortion in passband. The attenuation in rejected band can be corrected by the increase in the order of the filter. It was shown that if the upstream modulator is of order L, a cascade of K = L + 1 comb filters guarantees the necessary attenuation (Figure 7.13) [CAN 86]. Comb filters are often implemented in a recursive form [GAO 00]. In this case, if we suppose a data representation in addition to 2, in order to avoid the overload, the width of the registers must be at least $br = bin + K.\log_2(M)$, where bin is the number of data bits at the filter's input, K is the number of comb filters in cascade and M is the decimation factor (Figure 7.14) [ROG 81].

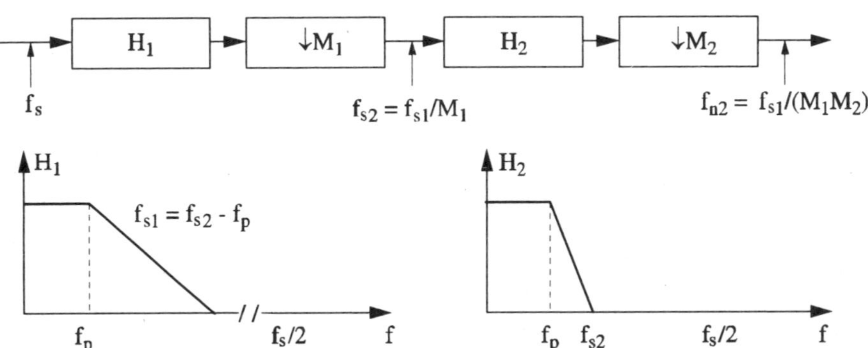

Figure 7.12. *Example of chain for filtering/decimation: the filter of the first stage undergoes a strong sampling rate, but has a large transition bandwidth of important transition. The filter of the second stage presents a narrow transition band, but benefits from a lower sampling rate. The constraints in terms of ratio (Δf/fe) for the two filters are thus toned down*

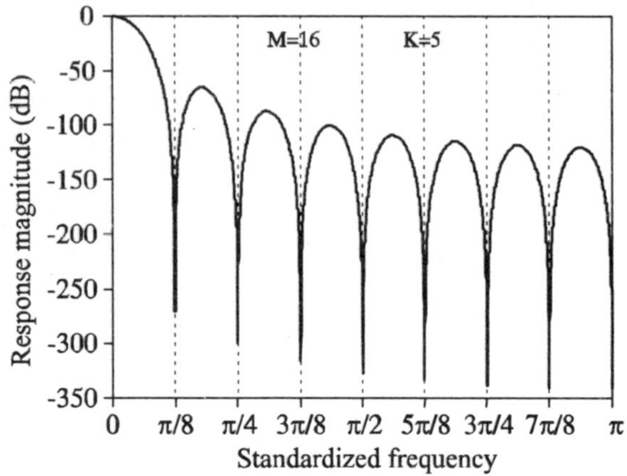

Figure 7.13. *Frequency response for a cascade of comb filters adapted to a modulator of order 4*

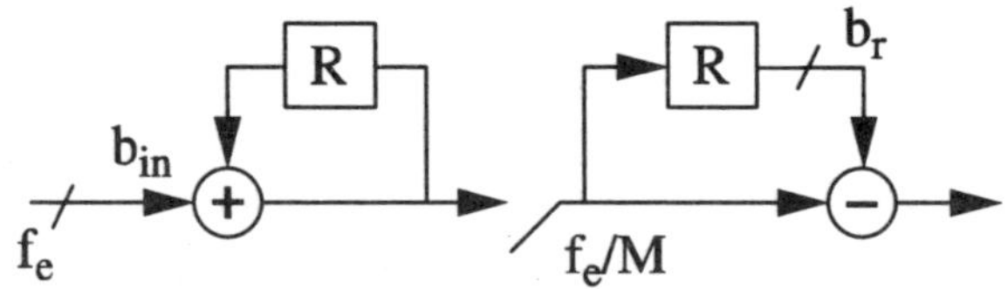

Figure 7.14. *Recursive implementation structure for a comb filter*

The compensation of the distortion in passband can be carried out by a correct filter. In order to limit this distortion, the comb's stopbands width must be sufficiently large in order to meet the specifications. The comb filters are used for a maximum decimation such that the filter output frequency corresponds to $4 \times fN$, where fN is the Nyquist frequency of the signal [CHU 84]. FIR (*finite impulse response*) or IIR (*infinite impulse response*) filters can be used for subsequent filtering. The symmetry of linear phase filters enables a 50% reduction in the number of multiplications. IIR filters, on the other hand, can be designed with an order lower than that of an equivalent FIR filter, but at the expense of a non-linear phase. Rather than correcting the non-linearity of the phase of the IIR filters, a possibility is the design of almost-linear phase IIR filters [LAW 97, KRU 99]. In [GRA 01], we can see an example of IIR filter installation for GSM and DECT. However, FIR type filtering remains more widely spread than IIR filtering due to the constraints in phase linearity imposed by certain applications. Among FIR filters, we can distinguish the filters called half-band, which are very well adapted to the decimation by a factor 2. The symmetry presented by the frequency response of the half-band filters makes that half of the coefficients equal to zero, thus making them particularly interesting from the point of view of the calculation complexity (Figure 7.15).

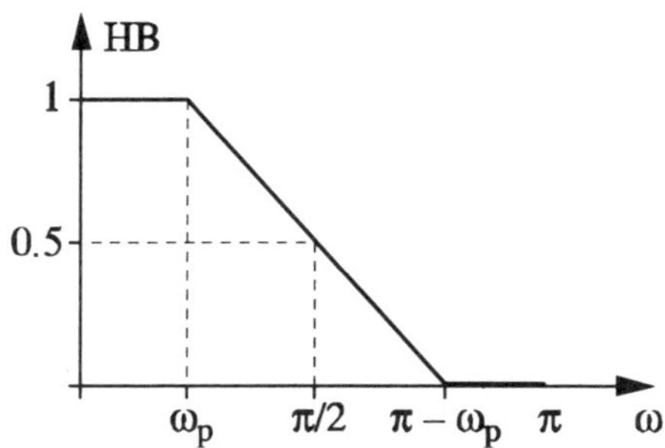

Figure 7.15. *Frequency response of a half-band filter*

As far as the implementation of filter decimators is concerned, a current solution consists of a cascade of comb filters, which carries out a preliminary decimation, followed by a FIR filters cascade (half-band or not) which compensates the response of the comb filter and concludes the decimation process. In other words, the filters based on multipliers are placed at the end of the chain in order to benefit from a less significant sampling rate. However, the low number of bits at the output of ΣΔ modulators reduces the complexity of multipliers and there is a compromise to establish [NAV 00]. In conclusion, after choosing the filter response/specification, an optimization process is essential in order to take into account the properties of the filter's response which have an impact on the performance and implementation cost. Apart from the symmetry, other parameters such as the effective number and the minimal precision of coefficients and characteristics of the target technology will make it possible to establish a more realistic model of hardware requirements and the consequences in consumption for a given implementation (ASIC, DSP, FPGA, etc.).

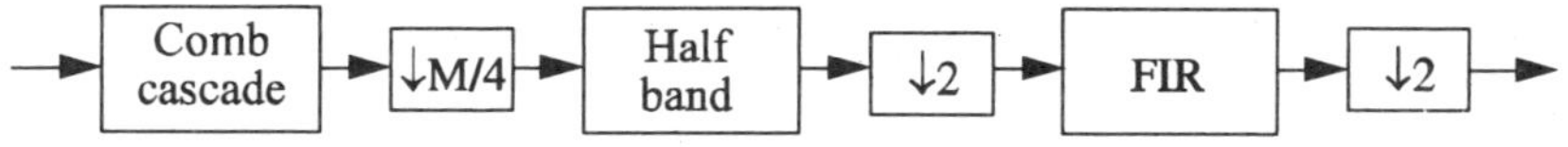

Figure 7.16. *Example of chain for filtering chain and decimation by M factor*

7.5. ADC evolution

Various sources make it possible to have an idea of the evolution of ADC performances in time. The most interesting source is the ITRS (*International Technology Roadmap for Semiconductors*).

The latest data dating from 2005 are presented in Figure 7.17. They present the evolution of the *FoM* merit factor according to the years. This factor is defined by the following relation:

$$\mathrm{FoM}_{\mathrm{CAN}}\left[\mathrm{GHz/W}\right]10^{3}=\frac{2^{\mathrm{ENOB}}\cdot\min\left(\left\{F_{s},2B\right\}\right)}{P} \qquad [7.9]$$

where ENOB is the effective number of bits, F_S the sampling rate, B the useful band of the signal and P the power dissipation. This graph provides the following comments:

– currently this merit figure is approximately 10^3 GHz/W. If we set a resolution of 10 bits and a frequency of 1 GHz, the consumption must be 1 W;

– if we reduce the resolution with a bit, we can divide the consumption with a 2 factor;

– there is a high long-term uncertainty. If we suppose that this factor will reach the value of 10^4 GHz/W in about 15 years, this will enable us to obtain, for example, a converter: 10 bits, 1 GHz, 100 mW.

The preceding data assumes the circuit is at room temperature. An important research activity in superconductor electronics already provides very good experimental results [BUL 02]. The sampling speed reaches 40 GHz and we demonstrated the creation of a passband sigma delta modulator centered on a frequency of 2.23 GHz. The noise measurement in a 19.6 MHz band of is of – 57 dB with respect to the full scale. The modulator consumes only 1.9 mW. The difficulty is for the temperature: 4.2°K. This type of converter can reach the performances expected by the needs of the software radio. However, for the moment this solution is limited by the cost and the consumption of the cryogenic environment.

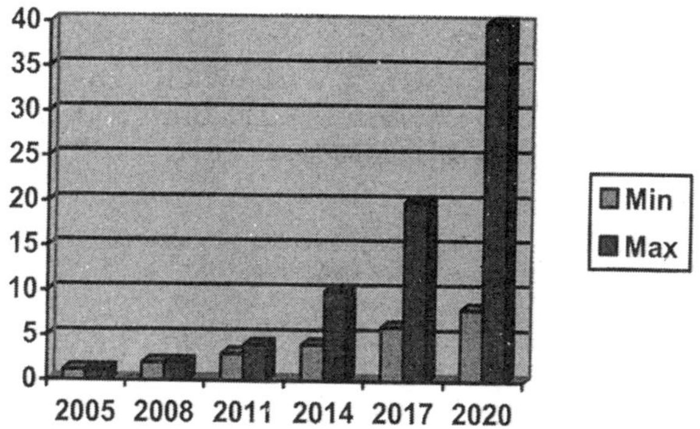

Figure 7.17. *Merit factor evolution in time from 2005 to 2020 (source: ITRS)*

7.6. Conclusion

The ideal software radio imposes specifications on the ADC in the receiver which cannot be reached in the next 15 years in the evolution of semiconductor technologies. Alternative and suboptimal solutions such as signal sampling architecture in intermediate frequency or zero intermediate frequency make it possible to tone down the constraints on the converter.

Currently, the two most widely used ADC architectures in the application concerned are the pipelines and the sigma delta. In the future, possible innovations on these techniques could significantly contribute to the evolution of performances in speed, resolution and power dissipation. The parallelization of these structures is a complementary way to follow.

Other technologies such as superconductors, which are too expensive for the moment, but very promising in the future, are explored.

The future requirements in radiocommunications, particularly for the reconfigurable systems, impose significant research on receiving transmitters and particularly on ADCs. The performances of the latter will significantly evolve in the next 10 years.

7.7. Bibliography

[BUL 02] BULZACCHELLI J.F. LEE H., MISEWICH J.A., KETCHEN M.B., "Superconducting Bandpass ΣΔ Modulator with 2.23-GHz Center Frequency and 42.6-GHz Sampling Rate", *IEEE Journal of Solid-State Circuits*, vol. 37, no. 12, p. 1695-1702, December 2002.

[CAN 86] CANDY J.C., "Decimation for Sigma-Delta Modulation", *IEEE Transactions on Communications*, vol. COM-34, p. 72-76, 1986.

[CHE 00] CHERRY J.A., SNELGROVE W.M., *Continuous-Time Delta-Sigma Modulators for High-Speed A/D Conversion – Theory, Practice and Fundamental Performance Limits*, Kluwer Academic Publishing, 2000.

[CHU 84] CHU S., SIDNEY BURRUS C., "Multirate filter designs using comb filters", *IEEE Trans. Circuits and Sys.*, vol. CAS-31, p. 913-924, November 1984.

[CRO 83] CROCHIERE R.E., RABINER L.R., *Multirate Digital Signal Processing*, Prentice-Hall, 1983.

[ELW 01] ELWAN H., ALZAHER H., ISMAIL M., "A New Generation of Global Wireless Compatibility", *IEEE Circuits and Devices Magazine*, no. 17, p. 7-19, January 2001.

[GAO 00] GAO Y., TENHUNEN H., "A fifth-order comb decimation filter for multi-standard transceiver applications", *Proceedings of ISCAS 2000 – IEEE International Symposium on Circuits and Systems*, Geneva, 28-31 May 2000.

[GIN 92] GINETTI B., JESPERS P.G.A., VANDEMEULEBROECKE A., "A CMOS 13-b Cyclic RSD A/D Converter", *IEEE Journal of Solid-State Circuits*, vol. 27, no. 7, p. 957-964, July 1992.

[GOE 94] GOES J., FRANCA J., PAULINO N., GRILO J., TEMES G., "High-Linearity Calibration of Low-Resolution Digital-to-Analog Converters", *IEEE Int. Symposium on Circuits and Systems*, 1994, p. 345-348.

[GRA 01] GRATTI K., GHAZEL A., NAVINER L., TABBANE S., "Comparison of FIR and IIR Structures for Decimation Filtering in Radio Communications", *Proc. of the 5th Multi-Conf. on Systemics, Cyb. and Informatics*, SCI'2001, Orlando, USA, July 2001.

[KRU 99] KRUKOWSKI A., KALE I., "Almost Linear-Phase Polyphase IIR Lowpass/highpass Filter Approach", *International Symposium on Signal Processing and its Applications (ISSPA'99)*, Brisbane, Australia, August 1999.

[LAW 97] LAWSON S.S., "On Design Techniques for Approximately Linear Phase Recursive Digital Filters", *IEEE International Symposium on Circuits an Systems*, Hong Kong, June 1997.

[LEE 92] LEE S.H., SONG B.S., "Digital-Domain Calibration of Multistep Analog-to-Digital Converters", *IEEE Journal of Solid-State Circuits*, vol. 27, no. 12, p. 1679-1688, December 1992.

[LEW 87] LEWIS S.H., GRAY P.R., "A Pipelined 5-Msample/s 9-bit Analog-to-Digital Converter", *IEEE Journal of Solid-State Circuits*, vol. 22, no. 6, p. 954-961, December 1987.

[LEW 92] LEWIS S.H., SCOTT FETTERMAN H., GROSS G.F., RAMACHANDRAN R., VISWANATHAN T.R., "A 10-b 20-Msample/s Analog-to-Digital Converter", *IEEE Journal of Solid-State Circuits*, vol. 27, no. 3, March 1992.

[LOU 02] LOUMEAU P., NAVINER J.F., PETIT H., NAVINER L., DESGREYS P., "Analog to digital conversion: technical aspects", *Annals of Telecommunications*, vol. 57, no. 5-6, May-June 2002, p. 338-385.

[MIN 00] MING J., LEWIS S.H., "An 8b 80MSample/s pipelined ADC with background calibration", *IEEE International Solid-State Circuits Conference*, Digest of Technical Papers, p. 42-43, 446, February 2000.

[MIT 95] MITOLA J., "The Software Radio Architecture", *IEEE Communications Magazine*, vol. 33, no. 5, p. 26-38, May 1995.

[NAV 00] NAVINER L., NAVINER J.F., "On efficient cascade implementation of narrow band decimator filter for sigma delta modulators", *Proceedings of the IEEE Midwest Symposium on Circuits and Systems*, MWSCAS2000, Lansing, USA, August 2000.

[RAZ 97] RAZAVI B., "Design Considerations for Direct-Conversion Receivers", *IEEE Transactions on Circuits and Systems-II*, vol. 44, no. 6, June 1997.

[RAZ 03] RAZAVI B., "RF CMOS Transceivers for Cellular Telephony", *IEEE Communications Magazine*, August 2003.

[ROG 81] ROGENAUER E., "An economical class of digital filters for decimation and interpolation", *IEEE Trans. on Acoustics, Speech, and Signal Processing*, April 1981.

[RUD 97] RUDELL J. *et al.*, "A 1,9 GHz Wide Band IF Double Conversion CMOS Integrated Receiver for Cordless Telephone Applications", *IEEE Journal of Solid-State Circuits*, vol. 32, no. 12, p. 2071-2088, December 1997.

[SOE 95] SOENEN E.G., GEIGER R., "An Architecture and an Algorithm for Fully Digital Correction of Monolithic Pipelined ADCs", *IEEE Transactions On Circuits and Systems II*, vol. 42, no. 3, p. 143-153, March 1995.

[WAL 99] WALDEN R.H., "Analog-to-Digital Converter Survey and Analysis", *IEEE Journal on Selected Areas in Communications,* vol. 17, no. 4, p. 539-550, April 1999.

Chapter 8

Flexible Spectrum Management

8.1. Introduction

All wireless communication systems – such as commercial mobile radio systems (e.g. GSM), private systems (e.g. civil security) or radio broadcasting (e.g. television) – use the radio spectrum (or more improperly the radio) as a physical medium for information transmission. The inability to create radio equipment that eliminates interference has artificially led to the radio resource[1] being considered as limited. Consequently, controllers have long established static and rigid rules for spectrum use, with the only goal of better managing interference in radio systems. In this light, each radio system has been designed for a given application and inside a band dedicated exclusively to its operation. Taking into account the limited spectral band dedicated to each system, spectrum sharing techniques have been created in order to enable a better use (spectral efficiency) of the spectrum. These techniques deal with the physical aspects of a transmission (modulation, coding, signal processing, etc.) and with sharing among several transmissions (spatial, temporal and spatial multiplexing techniques, management algorithm of radio resource among the cells, etc.). These techniques have been developed inside the same system. In this chapter, spectrum sharing is no longer considered within a single system but between several systems.

Chapter written by David GRANDBLAISE.

1 In the rest of this chapter, the terms "radio resource"; "spectral resource" and "spectrum" will be indifferently used in order to designate the spectral resource.

The recent onset of the desire to make the spectrum control rules more flexible, on the one hand, and to introduce the software defined radio SDR at the level of radio equipment (thus considered reconfigurable), on the other hand, offers new perspectives for the management of radio resources. Notably, this makes it possible to consider a more flexible spectrum management inside the same or several radio systems. This chapter emphasizes these new scenarios of flexible spectrum management while observing the regulation and the latest technological progresses.

The second section presents the flexible spectrum management drivers. The new concepts and models of flexible spectrum management are introduced and presented in the third section. This section presents the different and current thoughts and proposals on this subject expressed by different regulatory authorities (national and worldwide) or by research groups. The implementation of advanced scenarios for the flexible spectrum management requires advanced technologies which are complementary to SDR. The fourth section presents the principles of the most promising technological facilitators, such as "cognitive radio", which make it possible to implement the most advanced concepts of flexible spectrum management.

8.2. Flexible spectrum management drivers

The reconsideration of the current practice of spectral resource management in the telecommunication systems has been initially raised in the USA. These drivers are both economically and technically originated. These drivers are described and debated in this section.

8.2.1. *The spectrum is not rare, it is used inappropriately*

Until nowadays, the spectrum allocation to a technology and the frequency allocation to an operator are managed by the international (e.g. ITU, CEPT) and national (e.g. ART, ANFR in France) regulatory bodies. One of the main roles of these bodies is to ensure the spectral co-channels and adjacent channels coexistence between radio systems operating within the same band or adjacent bands. This practice for co-channels interference management and adjacent channels has been there since the beginning of the regulation for radiocommunication systems. This is mainly due to the fact that the radio systems have not had until now the capacity to efficiently distinguish the necessary signal from the interfering signal when more systems operate in the same band or in very close bands. Hence, this regulatory practice, which is based on the precaution principle for interference management, has allocated and has assigned to each system (and therefore to each operator in case of licensed systems) a clearly defined and identified spectrum quantity inside the radio electrical spectrum. The different systems, as the different carrier frequencies inside a system are separated by guard bands in order to control *a priori* the interferences. By examining these practices a little closer, the sections below will show that this practice is far from being optimal in terms of spectral efficiency.

8.2.2. *Spectrum reuse, connection opportunities and reconfigurable radio equipment*

The improvement of spectral efficiency depends particularly on the capacity to reuse the spectrum in space. This section analyzes the relation between the need of spectrum quantity and the density of access points which are necessary to support a given traffic inside a given geographical area.

Whatever the considered radiocommunication system (cellular, broadcasting, WLAN), the assignment method of frequencies has been specifically defined for each of these systems under the hypothesis that these systems are used independently from each other inside their respective spectral band initially allocated to them. Each of these bands is static and limited in terms of number of frequencies. However, this initially limited number of available frequencies affects the frequency[2] spatial reuse and consequently affects the necessary number of radio access points (TRX[3]) in order to carry a given traffic demand. Therefore, this means that the need in spectrum (noted by bw) is closely linked to the access point density (noted by ε). In other words, the higher ε is, the easier the frequency reuse becomes and thus weaker bw can be. Thus, the lack of spectrum problem is a false problem. It is more a problem [MIT 99a] of lack of connection opportunities at the radio access points (therefore linked to ε). In fact, for a fixed bw, there is no theoretical limit to indefinitely increase the capacity when ε increases [ZAN 97]. In practice, ε is limited by the limited number of sites which can be used by an operator.

2 The frequency spatial reuse is the physical spatial distance (geometrical meaning) from which two transmitters can transmit without mutual interference and using the same frequency.

3 TRX = *transceiver* = a couple (transmitter-receptor) to support a frequency of a given radio access technology.

As an example, in the case of GSM (capacity limited system), the relation (compromise) between ε (TRX/km^2) and bw (MHz) can be mathematically expressed for a hexagonal model in the expressions below. The variables used are:

– ud denotes the user density (user number/km^2);

– ut expresses the intensity of the traffic per user (Erlang per user);

– ErlgB^{-1} (inverse of the Erlang B law) denotes the necessary number of channels in order to support a given traffic and blocking probability $P_{blocking}$;

– K is the cluster size; w is the bandwidth for a GSM carrier;

– N_s is the number of *slots* for each frame;

– γ characterizes the propagation environment;

– nbc is the number of duplex channels;

– C/I is the carrier signal-to-interference ratio;

– R is the cell radius.

Expression [8.2] below shows that the smaller R is, the bigger ε and the smaller bw. This expression does not linearly vary with ε. The offered capacity ac (Erlg/km^2/MHz) by link (ascending or descending way) can be easily expressed through:

$$ac = \frac{2 \bullet ud \bullet ut}{bw} \qquad [8.1]$$

Equation [8.1] shows that ac is inversely proportional to bw and thus proportional to ε.

$$bw = K \bullet 2 \bullet nbc \bullet w \text{ with } K = \frac{1}{3} \bullet \left(6 \bullet \frac{C}{I}\right)^{\frac{2}{\gamma}}, \ \gamma = 3 \text{ and } w = 0.2 \text{ MHz}$$

$$bw = K \bullet 2 \bullet \frac{Er\lg B^{-1}(P_{blocking}, A_{cell})}{N_s} \bullet w$$

with $N_s = 8$ slots, $P_{blocking} = 2\%$ and A_{cell} the traffic offered by the cell;

$$bw = K \bullet 2 \bullet \frac{Er\lg B^{-1}(P_{blocking}, ud \bullet S_{cell} \bullet ut)}{N_s} \bullet w$$

with S_{cell} the surface of a hexagonal cell;

$$bw = K \bullet 2 \bullet \frac{Er\lg B^{-1}(P_{blocking}, ud \bullet \frac{1}{\varepsilon} \bullet ut)}{N_s} \bullet w \quad \text{with } \varepsilon = \frac{2}{3\sqrt{3}R^2} \qquad [8.2]$$

The spectrum is therefore not rare, but the connection opportunities with the access points are, leading to an under optimal use of the frequency reuse distance. Considering this property inside a multi-operator context and a reconfigurable radio equipment (based on the SDR technology), it is possible to create a greater flexibility for radio resource sharing. In fact, if, for example, the cell base stations are reconfigurable (multimode and multiband in the same time), each base station can in principle support the connection of any radio access technology with an SDR terminal (single mode or multiband). This reconfigurable equipment thus makes it possible to increase the possibilities to reuse the frequency because a terminal can easily find a closer access point (by supposing here the ideal case where each access point can support any radio interface, whereas the approach is probably different in practice).

8.2.3. *Sporadic use of spectrum in time and space*

The capacity to reuse the spectrum in a more efficient way such as described in section 8.2.2 is a radio engineering issue. Beyond these purely engineering aspects, the spectrum use measurement campaigns have shown in addition that the spectrum is also under-used by different radio access technologies, opening new opportunities for a second use of the spectrum.

The under-use of the spectrum can be analyzed from two perspectives: temporally and spatially. As an example, the right part in Figure 8.1 shows the traffic fluctuations (and therefore of the spectrum assuming the simplifying hypothesis that traffic spectrum demand and traffic intensity are proportional) for different services within time-space plane.

For a given spatial localization, this figure shows the traffic temporal variation. In the same manner, for a given time instant, the spectrum use is also geographically variable. Generally, the cellular networks deployment (in terms of number of sites /deployed TRX) is sized for the rush hour (i.e. the peak of traffic). However, this worst case corresponding to the complete use of the spectrum is real only during a time fraction. Hence, the spectrum is under used during the rest of the time. Significant temporal variations have been observed for the voice service inside the cellular radio systems [ALM 99, LAM 97], but also for radio TV broadcasting system [KIE 98] and audio [RAD 05]. These observations also show that each service has its own temporal profile of the traffic. These profiles present a certain temporal decorrelation. This decorrelation makes it possible to consider that certain systems with spectrum punctual demand can temporarily reuse the spectrum of those that partially use theirs. A more complete analysis [MIT 00] of the traffic variation shows how the traffic temporal variation (e.g. morning, afternoon, night, weekend) is tightly linked to the traffic spatial variation (e.g. airport, urban, suburban, residential environment) for different services (e.g. public security, commercial systems). The surface of geographical areas where any temporal variation happens corresponds to several cells, i.e. it includes a certain number of users inside a geographical area where the traffic gradient is relatively constant during this period.

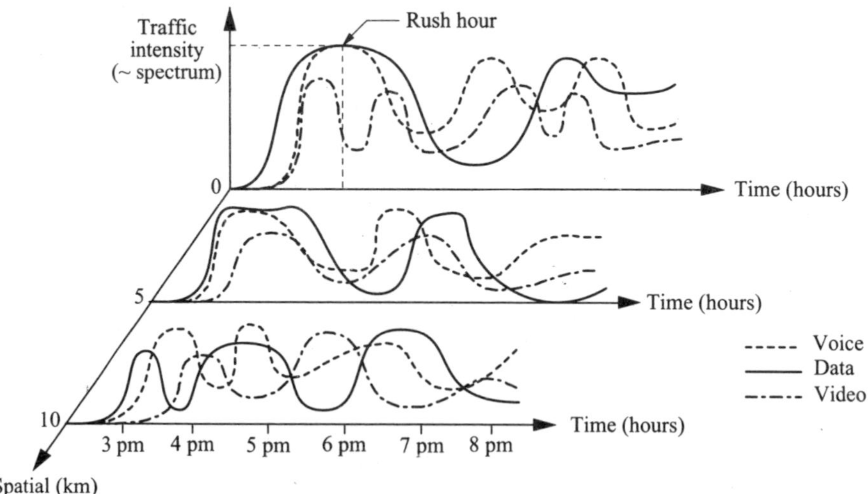

Figure 8.1. *Variable use of spectrum in time and space*

For the civil and military applications, other measurement campaigns have shown that the spectrum use also has inactivity periods (spectral holes) for minutes or seconds, even milliseconds within localized geographical areas. The measurements have particularly shown that the probability that at least 4 MHz of the band should be available during one second is of 95% [DUP 03]. For this very short time scale, the rate of the activity of the spectrum can be (Figure 8.2[4]): *interruptive* (the spectrum is intermittently used during periods of a few seconds), or fully *dynamic* (the spectrum is intermittently used during periods of milliseconds). This sporadic inactivity of the spectrum is different depending on the place where the measurement is taken. Therefore, at every instant, each user (terminal) or each radio access point (for example, base station) located in different places perceives spectrum use with generally spectrum holes to their radio environment (in relation to propagation conditions, to the presence or non-presence of close interferences). In other words, the opportunities of spectrum reuse correspond to each "space-time" point.

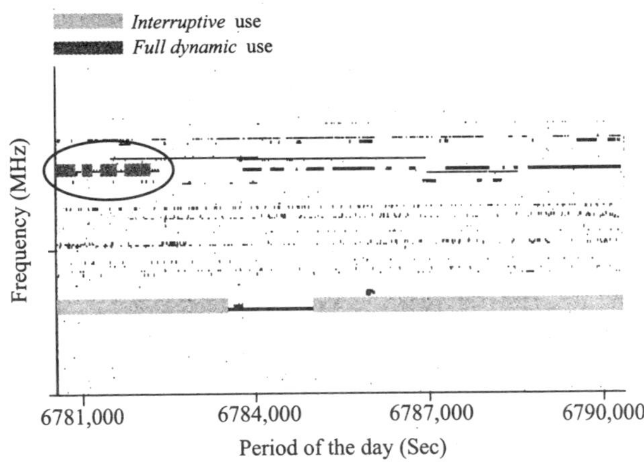

Figure 8.2. *Sporadic use of the spectrum*

8.2.4. *The opportunities for flexible spectrum management*

The previous observations have shown that the under use of the spectrum is different in time and space and for different resolutions of these two dimensions. By

4 Adapted from the presentation made by the DARPA, The USA, Industry Journal, October 2001.

generalizing these observations, temporal and spatial resolutions can be ranked as follows.

Time resolution:

– very short duration (ms, s, min.). This corresponds to the previous *interruptive* and *fully dynamic* features;

– medium duration (10 minutes, half an hour, an hour);

– long duration (half a day, a day, a week);

– very long duration (a week, a month, seasons).

Spatial resolution (coverage):

– very weak coverage (local, user-centered);

– average coverage (*hot spots, cell*);

– large coverage (urban, suburban areas);

– very large coverage (rural area).

Created on this terminology, the opportunities plane for the reuse of the spectrum is presented in Figure 8.3.

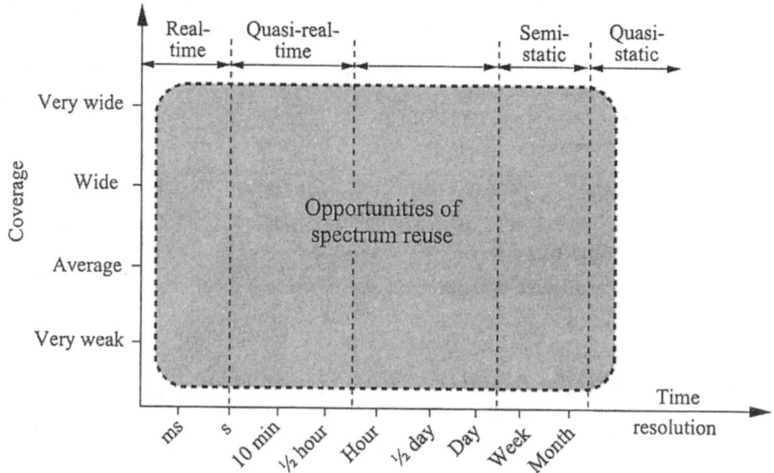

Figure 8.3. *Time-space plan for spectrum reuse*

For very short time scales (real-time), the spectral holes can be used for very short transmissions, i.e. pieces of information of the size of one or more packets. For quasi-real-time scales, the opportunities of spectrum reuse are close to the average time of a person's communication and therefore the time period of a session. Hence, the real-time and quasi-real-time scales are appropriated for a radio link use. Due to longer variations rhythms, the real semi-time, semi-static and quasi-static time scales involve a greater number of connections, i.e. radio links. Consequently, the reuse opportunities for this spectrum are carried on more significant band quantity, i.e. on multiple spectrum quantities of a system's bandwidth (e.g. more 200 KHz carriers in a GSM system).

From these observations, it seems that the nature of the opportunities of spectrum reuse depends on the rhythm of fluctuations of the spectrum use, i.e. depending on the considered time constant. Nevertheless, it is worth emphasizing that the opportunities of previous reuses (for one or more radio links) are not exclusive and they can even be considered conjointly and simultaneously. On the other hand, because the time constants are not of the same order, the techniques used to utilize these different types of opportunities do not necessarily require the same telecommunication functionalities.

8.2.5. *Resource sharing and economic impact*

Spectrum sharing

The strict rules imposed by the traditional regulation for a fixed and static allocation of the spectrum, i.e. an exclusive utilization of the spectrum by a given operator using a given radio access technology, has obviously constrained the *business model* developed by the operators in order to provide their services. This practice has led each operator to accept all the costs (CAPEX[5] and OPEX[6]) necessary for the network installation and maintenance. Particularly, these costs include the purchase of the radio license, of the sites, of the radio equipment and core network.

In a multi-operator context, the observations in section 8.2.3 have especially shown that the under-use of the spectrum makes it possible to consider a secondary use of the spectrum among operators (see section 8.3 for more details on this model). In fact, a second utilization returns to pool the spectrum, which can be used

5 CAPEX: *CAPital EXpenditure.*
6 OPEX: *OPerational EXpenditure.*

by several users according to their own needs. Particularly, the spectrum can be divided either among different operators by using the same radio access technology or even among different radio access technologies from different operators. These new perspectives have a direct impact on existing business models which have been previously debated. The business model is affected in two main points: (1) since potentially each operator can temporarily rent a part of its under-used spectrum to other parties, he can benefit better from his investment in license purchase; (2) since each operator can temporarily acquire (after negotiation) an additional part of the spectrum, he can find a new income source either by accepting more users, or by improving the quality of service. The dynamic market techniques (e.g. brokerage techniques, the auctions) are considered in order to implement the spectrum sharing. The exchange strategies created on these market concepts are nowadays subject for research and are discussed by the regulatory authorities (see section 8.3).

Sites and equipment sharing

Section 8.2.2 has underlined the compromise between the spectrum quantity and the sites density required to accept a given traffic inside a given geographical area. Inside the cellular networks, the fixed and static allocation of the current spectrum established by regulation has led the operators to make the network denser in order to discharge the more and more significant traffic. This has particularly contributed to find new radio sites to deploy the new equipment (base stations), reaching more and more increased deployment costs. Moreover, these practices used independently by each operator have pushed the latter to deploy similar radio equipment (e.g. BTS GSM) on different sites or very closely together in order to provide similar or different services. Thus, from an economic point of view, this approach is not efficient because it involves sites and radio equipment duplication, thus forcing each operator to support all investments (CAPEX) as well as the maintenance of sites and equipment (OPEX). From a sanitary point of view, this also has as consequence the increase of radio pollution phenomena. On the other hand, it is more and more difficult to find new sites even if the densification is necessary and the deployment of new technology requires new sites. Taking into account these observations and the compromise between spectrum quantity/sites density in order to support a given traffic, new radio engineering perspectives are considered among operators. In fact, since it is outside the periods of heavy traffic, not only the spectrum which is under used but also the sites and the radio equipment; the sites and radio equipment sharing can be considered in conjunction with the spectrum sharing enabling the operators to temporarily use the additional sites and radio equipment. In the same manner as the spectrum sharing, the dynamic market techniques (e.g. brokerage techniques, auctions) can be considered in order to implement the negotiation for sites and radio equipment sharing among operators. Therefore, this new approach also offers to different operators the opportunity to optimize their investment return for sites and equipment.

It must also be mentioned that the propagation conditions for the large coverage (therefore requiring smaller sites) are more favorable for the low bands. As an example (Figure 8.4[7]), a study has shown [SPE 03] how much CAPEX and OPEX a system can save if it is brought to no longer operate inside the 2GHz band but inside the radio broadcasting band (500 MHz). So, not only can the spectrum quantity be negotiated but also the nature of the spectrum to be utilized depending on the researched physical properties. This has a direct economic impact.

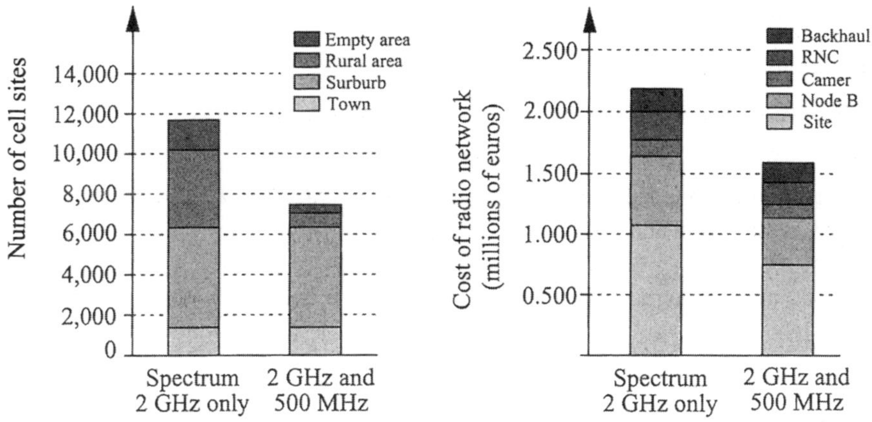

Note: costs and economies would occur in the operating expenses

Figure 8.4. *Example of CAPEX and OPEX economy during the use of low band (500 MHz) by a cellular system initially used inside the 2 GHz band*

This section has underlined the economic impact of resource sharing (spectrum, sites, radio equipment) among different operators in the case of cellular systems. These new considerations are important drivers motivating a more flexible spectrum management. Not only do they enable the operators to make their business model evolve (CAPEX and OPEX cost sharing among operators, dynamic management of capacity, QoS, new services) but they also enable the spectrum regulatory authorities to ensure that the spectrum is wisely utilized (if the spectrum is submitted to the market law in a dynamic manner, the spectrum is used by those who best use

7 Figure taken from [SPE 03].

it for the community) and that the spectrum efficiency is optimized (the spectrum is not under used). These considerations for a more flexible management of the spectrum will be even more significant once the base stations and reconfigurable SDR terminals are introduced. In fact, each base station will be capable to support any radio interference, thus increasing the connection opportunities for each terminal and this inside different bands.

8.3. Flexible spectrum management models

The ways the spectrum can be managed depend on the choice of the strategies considered for each of the following questions:

(a) which type of utilization must be authorized for the spectrum (spectrum allocation)?

(b) who must be authorized to use the spectrum (spectrum assignation)?

(c) how must the decision making be shared (centralized, distributed?) between the State on the one hand and the spectrum users on the other hand concerning the spectrum allocation and assignation?

Four main spectrum management models have been thus identified [FCC 02, EC 04]:

(1) *command and control* model;

(2) *commons* model;

(3) *market* model;

(4) *unrestricted usage* model.

While the traditional model (1) considered today does not offer any flexibility, options (2) to (4) are different models for a flexible spectrum management.

These four models are described and compared below. They are different in the way the decision making regarding allocation and assignation (discussed in (c)) are centralized and/or distributed between the State and the spectrum users. During this decision making, the primary and secondary allocation or assignation notions (for primary and secondary markets) are differentiated. It is worth mentioning that, in practice, it is possible to combine the various models.

8.3.1. *Command and control model*

This model [EC 04] represents the traditional approach of spectrum management. This refers to a static control whose rules [ITU 94] have been established by the International Telecommunication Union (ITU) for the entire world (divided in three areas). In this model, allocation and assignation management is the exclusive responsibility of regulatory authorities. The decisions regarding the spectrum allocation have a direct impact on the assignation and could lead to the exclusion of a certain number of candidate-users due to utilization restriction. In such a model, any possibility to change the spectrum allocation or assignation on a secondary market is impossible. This model assigns the license owner the right to exclusively use the frequency band. This license defines the service type which could be offered inside this band and also very often the restrictions on the radio technology type (e.g. maximal power level, standards) to be used in order to provide this service. In this model, the user has no formal decisional role for the band choice. Therefore, the international (e.g. ITU, CEPT) and national (e.g. ART) authorities are in charge of all the decisions and directly shape the market development. This static model has the advantage of avoiding all the situations of interference among systems (due to planning) both nationally and internationally. On the other hand, it transfers the responsibility of spectrum use to the regulatory bodies in terms of type of use, technologies and users. Hence, this model characterized by a long process does not have the necessary flexibility to respond to the market's rapid variations and to the introduction of innovative technologies. This is mainly due to the fact that the regulatory authorities, besides their role, have limited information regarding the real value brought over their choice (of users and utilizations). Nowadays, when the utilization demand is increasing, the number of technologies becomes larger and a spectrum need becomes more and more urgent to respond to the market's need, and the authorities' choice can thus become inappropriate very fast.

Due to these limitations, an interest for the new, more flexible spectrum management paradigms have sprung up. These forms consist mainly of reducing a certain number of utilization restrictions (for a more neutral use of the services and technologies), independently of the fact that utilization rights can be the subject of negotiations.

8.3.2. *Common model*

This model [EC 04] is the so called unlicensed spectrum model. This model is one of the first steps toward flexible spectrum management. This kind of model (e.g.

Wi-Fi) already exists. Within this model, the State is only in charge of the decision for the type of utilization regarding the service (i.e. the allocation) allowed for a specific band and also the type of technologies to be used within this band. On the other hand, this model enables an unlimited number (within the limits of the technical constrains) of users to share the access to the spectrum without previously having a license granted. With this model, the users do not have a formal protection against interferences by the establishment of regulation rules. However, protection is guaranteed by the implementation of standardized radio techniques and of appropriate radio etiquette[8] (sized so that the users use the spectrum only if they need it in order to optimize sharing). Even if today the *command and control* model is mainly used, there is an increasing desire to extend these *commons* models both for commercial systems (such as the point-to-point or point-multipoint broadband systems) and for the non-commercial ones.

8.3.3. *Market model*

In order to make the previous spectrum management models more flexible, tempestuous debates have recently taken place (and are still being debated) in order to find these new models [PEH 98]. A first stage toward these new models has been initiated by the economists [VAN 98, VAL 01, CAV 03] proposing a model where the operators can share the spectrum by using market mechanisms in order to acquire temporary rights for spectrum utilization. This model is also sometimes referred to in other works by the term *licensed open spectrum* [PEH 98].

Hence, the *market* model [EC 04] is considered a direct transition of the *command and control* model to a more flexible spectrum management. Within the *market* model the State is in charge only of the initial assignation of the user rights (primary users). After that, this model is based on the ideas that the market (and no longer the regulatory authorities such as the State) is fitter to determine which users (assignation) will make the most beneficial use (allocation) of the spectrum for society. This model therefore authorizes the transfer of spectrum utilization rights initially granted to the primary user toward one or more secondary users based on a secondary market. If in theory any right transfer can be considered within a neutral context of radio service and technology, in practice the establishment of technical rules enabling the spectral coexistence among systems should be necessary. This has the objective of protecting the users against unwanted interferences. These

8 By definition, the radio label is an assembly of RF bands, of air interfaces and of radio protocols, which moderates spectrum usage.

constraints can therefore limit the scenarios regarding the right sharing allowed by this model. The first estimations for capacity gains considered by this model have already been evaluated [LEA 04] for the continuous or fragmented spectrum sharing.

8.3.4. *Unrestricted usage model*

This model is one of the most flexible models among all the models presented and it is a radically different vision from the previous ones. In opposition to the previous approaches, this model is a completely uncoordinated (deregulating) version of the flexible spectrum management. This model is an extension of the previous *common* model toward a totally unlicensed model. This *unrestricted usage* approach is also referred to in other works under the name of *open spectrum* [NOA 95, WER 02, PEH 98, STA 04]. In opposition to the *licensed open spectrum* terminology used in section 8.3.3, this model can be named *unlicensed open spectrum.* Here, the regulatory authorities have no role in both the allocation and assignation procedures. This model offers the user a total freedom of deployment and type of service or technology if the interference is correctly managed with the bands which do not use this model. Consequently, the idea of primary and secondary systems does not exist in this model. This model cannot actually be realized with the existing radio technologies because they do not make it possible to grant sufficient protection against harmful interferences. Nevertheless, the introduction of new technologies (see section 8.4) should make it possible to put this model into practice in the near future. This model is defended by the US regulator FCC [FCC 02].

8.3.5. *Comparison of the models*

Table 8.1[9] compares the different previous models based on the allocation and assignation criteria centralized and distributed among primary and secondary systems.

---

9 Taken from [EC 04].

	Command and control **model**	***Common*** **model**	***Market*** **model**	***Unrestricted usage*** **model**
Primary allocation	*Centralized allocation* – the authorities are in charge of the allocation decision. The type of usage of the spectrum, the choice of associated services and technologies are predefined.	*Centralized allocation* – the authorities are in charge of the allocation decision. The type of usage of the spectrum, the choice of associated services and technologies are predefined.	*Distributed allocation* – the type of usage of the spectrum is decided by the market.	*Distributed allocation* – users have a total autonomy regarding the choice of spectrum type of usage.
Primary assignation	*Centralized assignation* – the regulatory authorities are in charge of the primary assignation of the rights for traditional spectrum utilization based on an exclusive utilization principle.	*Distributed assignation* – the access to the spectrum is not limited. An unlimited number of unlicensed users can use it. The users are not entitled to any regulatory protection against interferences.	*Joint centralized and distributed assignation* – the regulatory authorities are in charge of the primary assignation of the rights for traditional spectrum utilization based on an exclusive utilization principle. Yet, this is performed with market mechanisms such as auctions.	*Distributed assignation* – any user can use the spectrum. There is no regulatory authority to control or protect against interferences.
Secondary allocation and assignation	The secondary assignation (based on market mechanisms such as commission) is not authorized. The secondary allocation is not liberalized.	Not applicable because no right is transferable.	The secondary allocation is liberalized. The commission for the secondary assignation is authorized[10]	Not applicable because no right is transferable.

Table 8.1. *Comparison of spectrum management models*

10. This model is also sometimes referred to in other works as *licensed open spectrum* [PEH 98].

The case corresponding to the secondary allocation and assignation for a *market* model is already carried out in New Zealand especially for non-real-time exchanges. In Great Britain this specific model was the subject of advanced studies in order to estimate the impact of the introduction of a spectrum brokerage (for rights and obligations). In particular, this initiative has been started by Martin Cave [CAV 02] under the British regulator OFCOM. The British proposal is currently being evaluated in order to be put into practice in Great Britain. [OFC 04a, OFC 04b]. This proposal actually advises both the brokerage and the liberalization but also their combination.

In parallel, the Radio Spectrum Policy Group (RSPG) evaluates the opportunity to introduce the brokerage at European level in a harmonized manner among the Community countries [RSP 04a, RSP 04b]. From the international regulation point of view (ITU-R), some actions are also being initiated [WRC 03].

8.3.6. *Degrees of freedom and complexity*

Apart from the models of spectrum management previously described in section 8.3.5, Figure 8.5 presents a different analysis form these different models.

This figure underlines three big categories of spectrum management flexibility (increasing from left to right):

– *limited resource and absence of flexibility*. This case corresponds to the currently utilized models, which were previously described (*command and control* and *commons* models). The *command and control* model corresponds to a coordinated model, whereas the *commons* model corresponds to an uncoordinated model;

– *limited resources and partial flexibility*. This case corresponds to the previous *market* model, but can also include the hybrid *commons & market* model where an unlicensed system (secondary system) can aspire to use the licensed system spectrum (primary systems);

– *unlimited resources and total flexibility*. This case corresponds to the previous *unrestricted usage* model. Models (2) and (3) are part of the largest family of models, called *open spectrum*.

In the case of models called *open spectrum,* two modes for spectrum sharing are considered: the collaborative mode and the non-collaborative mode. In the collaborative mode, the secondary system informs the primary system regarding the intention to use its spectrum. This collaborative approach enables the primary system to rent its spectrum depending on a financial compensation. On the contrary, in the non-collaborative mode, the secondary system uses the primary system spectrum without previously informing the primary system, but with the warranty that it will not interfere. In the case of the hybrid model, the non-collaborative approach applied to unlicensed systems operating at levels close to the noise level (that of the licensed systems operating at high power) is also called *underlay*. An example of *underlay* is the *ultra-wide band* (UWB), recently authorized by FCC in 2002 to operate inside bands above 3 GHz. For each of the two modes, collaborative and non-collaborative, two strategies are considered: interruptible and uninterruptible. The interruptible approach authorizes the primary system to preempt all the secondary communications any time it needs the spectrum. This strategy is called *spectrum pooling*. The technical consequences of this strategy, especially at the MAC level, are described in section 8.4.4. On the contrary, the uninterruptible strategy does not authorize the primary system to preempt. In the case of the collaborative mode, the preemption right can be the object of an explicit mention within the contract concluded between the primary and secondary systems.

The models *market* or hybrid *commons and market* are a first stage towards a flexible spectrum management. Nevertheless, the flexibility brought by these models is based more on the introduction of new market mechanisms (which have not existed until nowadays) than on the technical innovation enabling a better management of the interference. On the other hand, the introduction of new technologies (see section 8.4) such as cognitive radio (see section 8.4.3) enabling better management of interferences will lead us to consider the spectrum as a quasi-unlimited resource (*unrestricted usage* model). The spectrum sharing mechanisms based on the market laws will no longer be necessary!

Thus, the market models can be considered as a short term solution for a flexible spectrum allocation, whereas the models based on a technology like the cognitive radio are longer term solutions. For sure, hybrid solutions will make the transition between these two approaches. These complementarities and transitions are presented in Figure 8.5; whereas the need for market mechanisms decreases, the need for cognitive radio functionalities increases from left to right.

It is important to mention that the flexibility models previously discussed are applicable for different time and spatial resolutions described in section 8.2.4. On the other hand, to each combination (model, resolution) corresponds a specific technical solution or solutions.

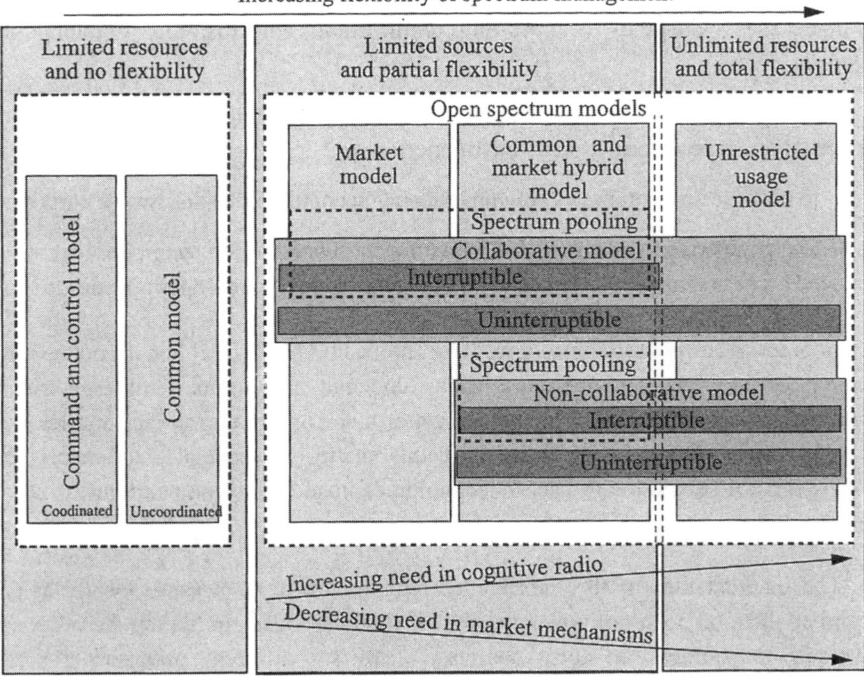

Figure 8.5. *Other representations of the models for flexible spectrum management*

8.4. The technologies

The existing radio systems have not been designed to naturally enable spectrum sharing following the different scenarios presented in section 8.3. Even if the technical constraints are different from a scenario to another, new technologies are necessary to move from the present practice (exclusive utilization of the spectrum by a system) toward a flexible management and spectrum sharing. This section presents the most pertinent technical concepts and technologies.

8.4.1. *Interference temperature*

The flexible spectrum management presented in section 8.3 supposes an elaborated coordination of the co-channel interference management among the same or different techniques of radio access. The management of this interference requires on the one hand *a priori* knowledge of the co-channel interference among technologies and on the other hand a characterization of the spectral holes in order to estimate the opportunity of spectrum reutilization. For this, the following are necessary:

– to characterize the co-channel interference knowing that different radio systems are not reciprocally sensitive to interference;

– to define a spectral hole knowing that each technology has its own sensitivity;

– to characterize a hole availability (a concept related to the technology we want to place here) in real-time in order to maximize the reutilization opportunities.

In order to respond to these objectives, the FCC[11] [FCC 03a] has introduced the concept of interference temperature whose objective is to define a universal metric enabling the characterization of these spectral holes by an appropriate interference measurement. The main application of this metric is especially to enable the coexistence of several radio access technologies inside the same band, at the same instant.

The inherent idea of the interference temperature is to measure locally at the recipient the total power of unwanted signal which has been emitted by the other co-channels transmitters or noise sources. There are different proposals for the definition of the existing interference temperature [FCC 03a]:

– expressed in degrees Kelvin (K), it can be calculated as the power (W) received by the antenna divided by the associated RF band (Hz) and Boltzmann's constant (W.s/K);

– it can also be calculated as the power density (e.g. $\mu W/m^2$ inside a given band) multiplied by the antenna's capture surface (m^2) and divided by an associated RF (Hz) and Boltzmann's constant (W.s/K);

– an interference temperature density (K/m^2) can also be defined as the interference temperature (K) per surface unit (m^2), i.e. the interference temperature (K) divided by the capture surface (m^2) of the reception antenna. This quantity could be measured for particular frequencies using a reference antenna, which makes it independent from the features of the reception antenna.

The utilization of this metric will enable assessment of the maximum admissible interference level. In other words, this metric will make it possible to characterize

11 Federal Communications Commission. The FCC is the US telecommunications regulatory body.

the worst radio environment where a recipient can operate. The inherent idea is to be able to specify this interference for a band, a geographical area and given services.

Figure 8.6 created by the FCC [FCC 02, FCC 03a] shows how the interference temperature can be used. This figure shows the necessary radio signal of a base station designed to operate at a maximal distance where the received power is close to the existing noise level at the time of system provisioning. While the new interfering signals arrive, the noise level locally increases consequently (represented by the peaks). This has as an effect on the coverage reduction. Based on this effect the interference temperature utilization can be useful at least for the two following applications: (1) within the licensed bands, it makes it possible to guarantee the admissible interference level. Any other noise sources which would exceed the threshold of the fixed interference temperature would be considered harmful. This would enable the sharer of this band to guarantee a more reliable use of the band; (2) as long as the threshold of the interference temperature is not exceeded, a second opportunistic use of the spectrum is possible by the other systems (e.g. the role of the unlicensed systems within the hybrid model). Hence, an interference temperature threshold could guarantee a maximum admissible interference level within this band.

For more detailed information regarding the utilization of interference temperature, see the FCC reports [FCC 02, FCC 03a].

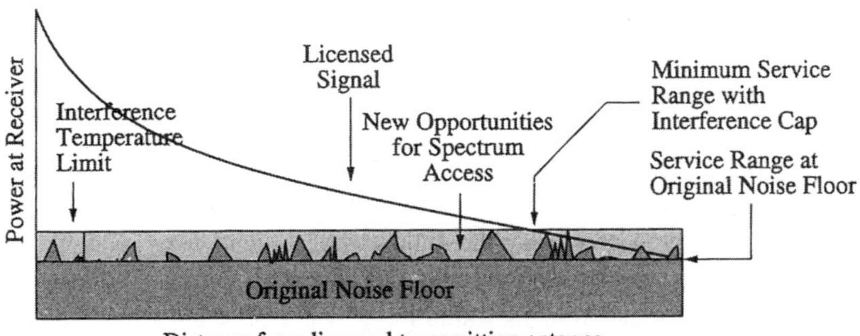

Figure 8.6. *Utilization of interference temperature metrics*[12]

12 Diagram taken from [FCC 03a].

8.4.2. *Forms of heteromorphic waves*

The forms of heteromorphic waves (called malleable) offer some interesting properties for a flexible sharing of the spectrum. These waveforms based on a multi-carrier technology have been recently proposed by the scientific community especially for a flexible sharing of the spectrum. Since the inactivity holes within the spectrum can be of a very narrow bandwidth and dissipated within the spectrum, a form of multi-carrier narrowband wave makes it possible to use all the forms of spectral holes. A first approach for this multi-carrier solution is to use a modulation OFDM [WEI 04] as multiple access technique. A second approach (very similar to the previous one) is to implement a multi-carrier solution based on the combination between multiple access techniques and interferometry techniques (CI) [HIJ 04]. Solutions such as CI/TDMA (*multi-carrier pulse shape*) [HIJ 04], CI/DS-CDMA (*multi-carrier chip shape*) [HIJ 04, MIC 03, FCC 03a], CI/MC-CDMA [PEZ 03, ZEK 03] and even FDMA/DS-CDMA [PEZ 03] are considered. In terms of relative performances [PEZ 03], the MC-CDMA presents a better spectral efficiency than the FDMA/DS-CDMA and a lower blocking rate in the case of an uninterruptive spectrum sharing.

There are a lot of reasons motivating the utilization of solutions based on the multi-carrier technique for flexible spectrum sharing. Moreover, it is possible to adapt the waveform (thus called heteromorphic) in a quasi-dynamic manner. This makes it especially possible to adapt the modulation type, the power control and also to gain diversity. Secondly, the multi-carrier technique makes it possible to use the bands unused by any radio access technique. In other words, each spectral hole corresponds approximately to a multiple of the minimum bandwidth of a multi-carrier system. Thirdly, the multi-carrier technique offers the possibility to use the band holes which can be continuous or discontinuous (fragmented).

These different properties make the multi-carrier systems very good candidates (1) for spectrum sharing among operators for licensed bands, (2) for spectrum sharing among unlicensed systems within the unlicensed bands (*unrestricted usage* model) and (3) to enable the unlicensed systems to use the licensed bands in an opportunistic manner.

In case (1), Figure 8.7 shows the spectrum sharing concept between two operators, both of them using the same multi-carrier technology. The (a), (b) and (c) parts in the figure show the case when the spectrum is not shared, the case when the spectrum is continuously shared and the case when a spectrum sharing can be done in a discontinuous manner (fragmented).

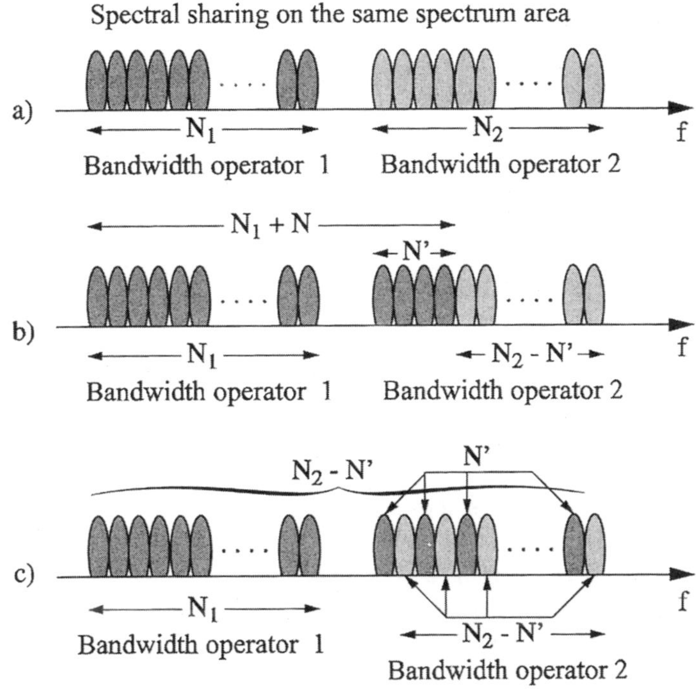

Figure 8.7. *Illustration of spectrum sharing among operators with a multi-carrier system for different strategies (no sharing, continuous sharing, discontinuous sharing)*[13]

Case (3) can take the form of *spectrum pooling* (see section 8.3). This strategy implementation called *spectrum pooling* [MIT 99a] can be approached using a multi-carrier solution OFDM [WEI 04] for the secondary system. A practical case [CAP 03] of *spectrum pooling* is the coexistence (*hotspot*) between a cellular system (e.g. GPRS) and a WLAN system (e.g. HiperLAN2) where the WLAN can aspire to reuse the GPRS band in an opportunistic manner. One of the major technical issues to consider is the secondary system capacity (i.e. WLAN) to detect during real-time the priority system's activity (i.e. GPRS) before this secondary system has started the transmission. The *spectrum pooling* has good performances for elastic traffic such as IP type traffic [CAP 03].

13 Figure borrowed from [HIJ 04].

8.4.3. *Cognitive radio*

The concept of *cognitive radio* introduced by Mitola [MIT 00] in 1999 is an extension of the software radio (SDR) leading to a tele-informatics[14] representation of the radio (meaning radio equipment). The idea introduced by this concept is to combine the local observations regarding the radio environment made by a radio equipment, knowing the available capacities of this radio equipment (*hardware* and *software*) and therefore to reach a decision regarding the way to modify the equipment behavior (i.e. its radio parameters) in order to reach the desired performance level. In other words, through cognitive radio, an equipment is capable to self-evaluate its *hardware/software* capacities, to self-identify the *hardware/software* constrains in order to support the constraints of a given service, to self-select the most appropriate waveforms, to find (i.e. to try, to diagnose) its RF environment appropriate to its situation, to decide and to operate in an independent manner (within the limits of its knowledge regarding the radio environment and its material capacities) in order to satisfy its own needs and not destroying the surrounding radio ecosystem. The application of this knowledge is based on a reasoning model [MIT 99b]. Therefore, cognitive radio can be considered as the control agent for software radio.

Apart from the introduction of the concept, Mitola [MIT 99a] has also widely established the principles and has proposed key-technologies of tele-informatics science enabling the implementation of cognitive radio. Cognitive radio architecture is represented by a cognitive cycle (Figure 8.8). It consists of several stages: (1) during the *observing* stage, based on external (e.g. RF environment) and internal stimuli (e.g. equipment battery level), a cognitive radio collects the group of stimuli which make it possible to recognize the radio context of the equipment and the user's communication needs; (2) the *orienting* stage establishes the emergency of the communication to be performed in order to create a priority order; (3) the *planning* stage has the objective to generate and to evaluate the alternatives (with or without network assistance) for the allocation of the radio resources most appropriate to the user need; (4) the *deciding* stage allocates the resources of computer processing to different software tasks (i.e. radio functions) enabling the application of radio resources allocation. For example, if the radio link becomes suddenly bad because of very damaged radio conditions, this stage is responsible for deciding the shift of this communication in a more appropriate radio frequency, i.e. with more favorable radio properties; (5) the *acting* stage starts the tasks using the dedicated resources and for a given duration.

---

14 Convergence between telecommunications and computer science.

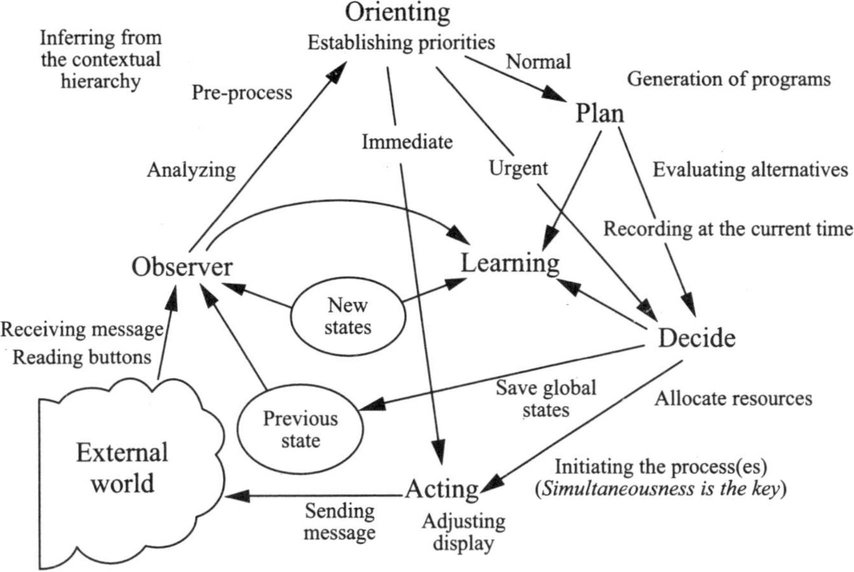

Figure 8.8. *Cognitive cycle*[15]

This cycle shows how the cognitive radio includes certain training process (generally assisted) in order to find out its errors. This cycle is very different from the existing cycle where the radio equipment exclusively and blindly considers the network instructions.

The interference temperature and the form of heteromorphic waves described in sections 8.4.1 and 8.4.2 are implicitly part of the cognitive radio principles and technologies. Hence, depending on the development of these principles and technologies, the cognitive radio provides advanced tools enabling the application of flexible spectrum management. The FCC [FCC 03b] has especially identified different application fields of the cognitive radio for the flexible spectrum management: (1) a licensed system can use the cognitive radio inside its own network that it operates in order to increase its own spectral efficiency; (2) the cognitive radio can facilitate the introduction of secondary spectrum market enabling agreements among operators and third parties (*market* model). In particular, in the future, the cognitive radio equipment will be able to directly negotiate the spectrum with the licensed system and to use it if and only if a

15 Figure taken from [MIT 99a].

previous agreement has been established between the cognitive radio equipment and the licensed system; (3) for different licensed systems providing the same services, the cognitive radio can facilitate the automatic frequencies coordination based on a cooperation. Such a collaborative coordination could rely on pre-defined rules previously established by the regulatory authority; (4) the cognitive radio could be used for a non-collective reuse (i.e. no previous negotiation) of the spectrum. For example, this would enable the unlicensed cognitive radio equipment to use in an opportunistic manner the licensed systems spectrum within geographical areas and when the latter do not use the spectrum (see the hybrid model in section 8.3.6).

In particular, a practical case for application (4) is to enable the unlicensed cognitive radio equipment to operate with high power within the rural areas. Based on the initial proposal of the FCC [FCC 04] to open the 54-698 MHz bands of the television band in the USA for a unlicensed usage, IEEE initiated (October 2004) the IEEE[16] 802.22 standard [IEE 04] enabling a deployment of a wireless regional network (WRAN[17]) based on the application of cognitive radio functions. The idea is to use in an opportunistic manner the spectral holes of the under used television band in order not to interfere with the licensed systems currently operating within this band. The objective of this standard is to design a cognitive radio interface (covering the physical layer and the MAC[18] protocol) operating within the 54-862 MHz band (i.e. within a band wider than that currently authorized by the FCC). In the end, the IEEE 802.22 standard should cover two types of scenarios:

– the use of the band by the low power radio equipment (up to 100 mW);

– the use of the band by high power fixed radio equipment which make it possible to provide broadband commercial services (point-multipoint).

In a similar way, the standard of the IEEE 802.16 group has recently (December 2004) suggested that the current IEEE 802.16-2004 radio interface (Wi-Max) could be extended in order to integrate cognitive radio functions (IEEE 802.16h) enabling two scenarios for flexible spectrum management:

– the coexistence between IEEE 802.16h systems themselves within license exempt bands;

– the coexistence between IEEE 802.16h systems and the current licensed systems (especially with the radio broadcasting system of the aforesaid television band, or 802.11 within 3.65-3.70 GHz band).

---

16 Institute of Electrical & Electronics Engineers.

17 *Wireless Regional Area Networks.*

18 *Medium Access Control.*

In light of the definitions of flexible spectrum management models presented in section 8.3, a question is raised: which are the fundamental differences between the traditional unlicensed Wi-Fi type (*commons* model) and the unlicensed model (*unrestricted usage* model) based on the cognitive radio? Below are presented some response elements. For example, contrary to Wi-Fi:

– the number of channels and the width of the channels are variable for a cognitive system;

– a cognitive radio system is capable of scanning the channel availability and managing the interference originating from other systems;

– a cognitive radio system can simultaneously transmit and scan the channel;

– for a cognitive system, the power level is not fixed, it can be adapted to the package level, etc.

Besides the emergence of these new activities toward the standardization of a first cognitive radio system, the interest shown for the study of advanced cognitive radio systems is the object of important research programs both in the USA [XG] and in Europe [E2R, MOE 03, MOE 04]. Working groups for cognitive radio have also been created inside the SDR Forum [SDR].

Intrinsically, the cognitive radio leads to an independent and distributed management of radio resources. Therefore, a major challenge is to model, analyze and size up such a radio system, called cognitive. The complexity of this task comes from the fact that the decision locally taken by a cognitive radio directly influences (and is influenced by) the decisions of other cognitive radios inside the system. With this effect, the study of cognitive radio systems can be addressed by using mathematical tools such as the game theory. The games theory presents the basic standards and the theoretical tools for (1) dealing with the conflict cooperation issue (e.g. contention problems with or without information sharing); (2) define cooperation strategies in order to reach a satisfactory balance for everyone. Applied to the field of cognitive radio [NEE 04], it makes it possible to especially manage the problems regarding the access control to the radio resource in the form of: (1) independent and distributed power control among the cognitive radio equipment [NEE 02]; (2) selection of the most appropriate waveform [NEE 02]; (3) application of spectrum sharing strategies based on the market theory (e.g. brokerage, auction) [WEB 98]. The application of this method to the cognitive radio is very promising. Nevertheless, significant research efforts are more than ever necessary within this field in order to better apprehend its potential.

8.4.4. *Cognitive radio etiquette*

The introduction of cognitive radio makes it possible to consider more flexible radio etiquettes. By definition, radio etiquettes are the set of RF bands, air interfaces and radio protocols moderated by the spectrum use.

Among the application fields of the cognitive radio for a flexible spectrum management, the FCC [FCC 03b] has particularly identified (see section 8.4.3) the cognitive radio as spectrum sharing facilitator based on a secondary market through agreements between the operators and third parties. Finally, the radio cognitive equipment will be able in particular to directly negotiate the spectrum with the licensed system and to use the spectrum if and only if a previous agreement has been concluded between the cognitive radio equipment and the licensed system.

Hence, the cognitive radio has to integrate automatic and independent reservation and access mechanisms into the spectrum which enable spectrum sharing in the form of real-time transactions (through the agreement or the contract based on a secondary market) between a licensed system and a third party. The cognitive radio must also integrate the mechanism, ensuring the return of radio resource to its owners when they need it. This is especially necessary when the spectrum sharing is based on an interrupting/preempting mode (e.g. *spectrum pooling*).

The implementation of these functionalities can be performed using the concept of a cognitive radio etiquette, i.e. a new type of MAC protocol. A first proposal for such a protocol has been proposed by Mitola [MIT 99a]. The general mechanism of this protocol is shown in Figure 8.9 for the case of spectrum discontinuous use. Here, the spectrum of a licensed system (*offeror*) can be used (based on a financial compensation) by secondary systems (*renter*) while the licensed spectrum is not used. Each line in Figure 8.9 represents a power level of the transmitted signal. The *offeror* starts the process by advertising the radio band to be rented through signaling. This signaling includes especially the information regarding the available frequency patch. The *offeror* listens for 10 ms if somebody is interested and then repeats the advertisement signal if necessary. A *renter* can show his interest for the location by using signaling in the form of encoded *express interest burst.* If the *offeror* hears this *burst*, the second *burst* specifies (*time, location*) the location time patch (beginning and end) and also the *price*. Following this proposal the *renter* submits a bid offer (*TLC bid*) and *authenticates* himself by using a short sequence. After this, the *renter* accepts/rejects the bid and then authenticates himself again. In the end, the process is completed by an *e-cash* prepayment done by the *renter.* During this stage, the *offeror* also acquires the necessary warranties to ensure the *renter's* solvability. If no objection is raised, the use of the spectrum by the renter is effective, otherwise the transaction is canceled. When the usage is finished the *renter* pays the rest of money.

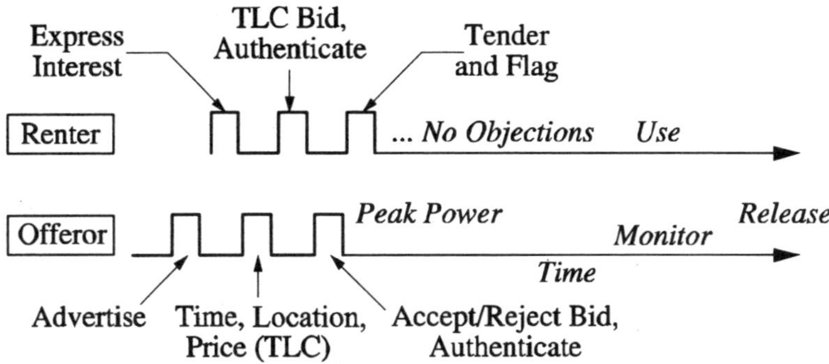

Figure 8.9. *Example of MAC protocol (spectrum pooling) enabling the negotiation for spectrum sharing between a licensed system (offeror) opening the spectrum use to a secondary system (renter)*[19]

If the spectrum owner prematurely needs the spectrum (e.g. civil security requirements) rented to a secondary party (*renter*), the owner has to use a mechanism enabling the reliable recovery (called reversion) of this spectrum. This can be done by using a *polite backoff protocol.* A possible solution (Figure 8.10) is to split the spectrum used by the *renter* into a periodical sequence consisting of two phases: (1) a phase (lasting 20 ms) containing the transmission of his data and (2) a listening phase (lasting 5 ms).

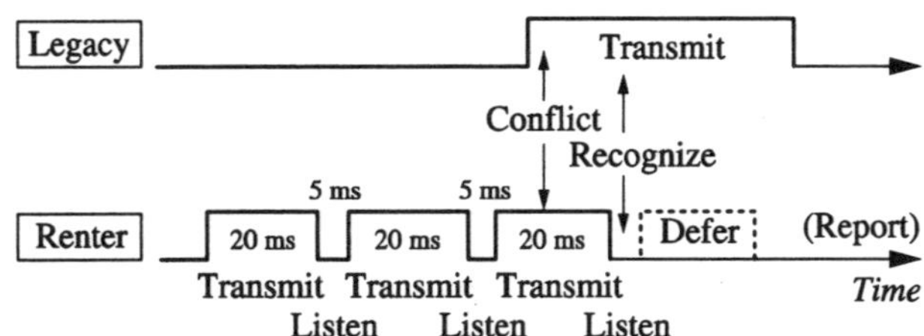

Figure 8.10. *Example of protocol recovery (called polite backoff) of the spectrum by its owner (legacy)*

19 Figure borrowed from [MIT 99a].

If the *renter's* signal level interferes with the *legacy* signal, there is a conflict. This conflict is identified by the *renter* during the listening phase. In this case, the radio channel is immediately released. An alternative channel for the *renter* is then sought and a financial compensation is offered to the *renter* because the channel was not available during the initial negotiated period.

8.5. Conclusion

The introduction of software defined radio at the radio equipment level offers a huge potential for the new perspectives regarding the flexible spectrum management at the system level. This revolution impacts on the present regulatory landscape whose rigid principles have served as cornerstones during almost a century in order to establish rules of spectrum allocation and assignation or design of radio technologies. These new perspectives have a direct impact on the current *business models* because the historical roles of regulatory authorities, operators, users and constructors are now completely overturned. In comparison with the current practice, this flexibility is twice as efficient. Not only that but, from an economic point of view, it makes it possible to ensure that the spectrum is used (in terms of value) within the society by those who "deserve" it most and operate it best, but it also guarantees that the spectral efficiency is improved from the technical point of view based mainly on the introduction of new intelligent technologies. The current thoughts of the research regarding the flexible spectrum management have been presented in this chapter. After having presented different drivers for a flexible spectrum management, the evolution of the current model of spectrum management towards more flexible models has been presented. It seems that there is not a single model for flexible spectrum management which will prevail but, on the contrary, it is very possible that in the future these different models will coexist or will be mixed. The potential of the new technologies, such as cognitive radio, have also been described. The research field for this subject becomes very active today. This field is more and more interesting since it is multidisciplinary, gathering at the same time the telecommunications, computer science, the cognitive sciences and the micro-economy.

8.6. Bibliography

[ALM 99] ALMEIDA S., QUEIJO J., CORREIA L.M., “Spatial and Temporal Traffic Distribution for GSM”, *VTC*, vol. 1, p. 131-135, 1999.

[CAP 03] CAPAR F., MARTOYO T., WEISS T., JONDRAL F., “Analysis of coexistence strategies for cellular and wireless local area networks”, *IEEE VTC'03*, Orlando, USA, November 2003.

[CAV 02] CAVE, M., *Review of Radio Spectrum Management*, UK Department of Trade and Industry, http://www.spectrumreview.radio.gov.uk, 2002.

[CAV 03] CAVE M., WEBB W., “Licence to interf”, *IEEE Communications Magazine*, December/January 2003/04.

[DUP 03] DUPONT, *Circle Spectrum Utilization during Peak Hours*, New America and Shared Spectrum Company, 2003.

[E2R] E^2R (End to End Reconfigurability) Project, IST FP6, http://www.e2r.motlabs.com.

[EC 04] ANALYSYS, ECON, HH, Final report for the European Commission – Study on conditions and options in introducing secondary trading of radio spectrum in the European Community, May 2004.

[FCC 02] FEDERAL COMMUNICATIONS COMMISSION, ET Docket no. 02-135, Spectrum Policy Task Force Group Report, November 15, 2002.

[FCC 03a] FEDERAL COMMUNICATIONS COMMISSION, ET Docket no. 03-108, Notice of Proposed Rule Making and Order, December 2003.

[FCC 03b] FEDERAL COMMUNICATIONS COMMISSION, ET Docket no. 03-322, Notice of Proposed Rule Making And Order, December 2003.

[FCC 04] FEDERAL COMMUNICATIONS COMMISSION, ET Docket no. 04-186, Notice of Proposed Rule Making And Order, December 2004.

[HIJ 04] HIJAZI S., NATARAJAN B., MICHELINI M., ZHIQIANG W., NASSAR C.R., “Flexible spectrum use and better coexistence at the physical layer of future wireless systems via a multicarrier platform”, *IEEE Wireless Communications Magazine*, vol. 11, no. 2, p. 64-71, April 2004.

[IEE 04] IEEE 802.22, site of standardization, http://www.ieee 802.org/22.

[ITU 94] INTERNATIONAL TELECOMMUNICATION UNION, *Radio regulations. 1990 Edition Revisited*, ITU, Geneva, 1994.

[KIE 98] KIEFL B., *What Will We Watch? A Forecast of TV Viewing Habits in 10 Years*, The Advertising Research Foundation, New York, 1998.

[LAM 97] LAM D., COX D.C., WIDOM J., “Teletraffic modeling for personal communications services”, *IEEE Communications Magazine*, vol. 35, no. 2, p. 79-87, 1997.

[LEA 04] LEAVES P., MOESSNER K., TAFAZOLLI R., GRANDBLAISE D., BOURSE D., TONJES R., BREVEGLIERI M., "Dynamic spectrum allocation in composite reconfigurable wireless networks", *IEEE Communications Magazine*, vol. 42, no. 5, p. 72-81, May 2004.

[MIC 03] MICHELINI, M., HIJAZI, S., NASSAR, C.R., ZHIQIANG W., "Spectral sharing across 2G-3G systems", *37th Asilomar Conference on Signals, Systems & Computers*, vol. 1, p. 13-17, 9-12 November 2003.

[MIT 99a] MITOLA, J., III, "Cognitive radio for flexible mobile multimedia communications", *Mobile Multimedia Communications (MoMuC '99)*, p. 3-10, 15-17 November 1999.

[MIT 99b] MITOLA, J., III, Cognitive Radio – Model Based Competence for Software Radios, Licentiate Thesis, Royal Institute of Technology (KTH), Stockholm, September 1999.

[MIT 00] MITOLA, J., III, Cognitive Radio – An integrated Agent Architecture for Software Defined Radio, Dissertation, Doctor of Technology, Royal Institute of Technology (KTH), Stockholm, May 2000.

[MOE 03] MOESSER K., GRANDBLAISE D., MOTTE N., CAPAR F., MOHYELDIN E.G., LUO J., "Techno-economic implications of end-to-end reconfigurability (E2R) systems", *10th Wireless World Research Forum*, New York, USA, October 2003.

[MOE 04] MOESSER K., LEAVES P., GRANDBLAISE D., CAPAR F., MOHYELDIN E., DOMESTICHAS P., DIMITRIKOPOULOS G., LUO J., TAFAZOLLI R., "Reconfiguration techniques to enhance network efficiency", *3rd Karlsruhe Workshop on Software Radio*, Karlsruhe, Germany, March 2004.

[NEE 02] NEEL J., BUEHRER R.M., REED B.H., GILLES R.P., "Game theoretic analysis of a network of cognitive radios", *IEEE MWSCAS-2002*, p. III-409 – III-412 vol. 3, 4-7 August 2002.

[NEE 04] NEEL J.O., REED J.H., GILLES R.P., "Convergence of cognitive radio networks", *IEEE WCNC*, vol. 4, p. 2250-2255, 21-25 March 2004.

[NOA 95] NOAM E.M., "Taking the Next Step Beyond Spectrum Auctions: Open Spectrum Access", *IEEE Communications Magazine*, vol. 33, no. 12, p. 6673, December 1995.

[OFC 04a] OFCOM, "Ensuring effective competition following the introduction of spectrum trading", consulted on 10th of June 2004, *http://www.ofcom.org.uk/consultations*.

[OFC 04b] OFCOM, "Notice of Ofcom's Proposals to Make Regulations: Spectrum Trading and the Wireless Telegraphy Register", consulted on 29 September 2004, *http://www.ofcom.org.uk/consultations*.

[PEH 98] PEHA J.M., "Spectrum Management Policy Options", *IEEE Communications Surveys*, vol. 1, no. 1, 4th trimester 1998.

[PEH 00] PEHA J.M., "Wireless Communications and Coexistence for Smart Environments", *IEEE Personal Communications Magazine*, vol. 7, no. 5, p. 66-68, October 2000.

[PEZ 03] PEZESHK A., ZEKAVAT S.A., "DS-CDMA vs. MC-CDMA, a performance survey in intervendor spectrum sharing environment", *The 37th Asilomar Conference on Signals, Systems & Computers*, vol. 1, no. 9-12, p. 459-464, November 2003.

[RAD 05] *Radio Today. How America Listens to Radio*, Arbriton Inc., New York, 2005.

[RSP 04a] RADIO SPECTRUM POLICY GROUP, "Public consultation on secondary trading of rights to use radio spectrum", http://rspg.groups.eu.int/consultations/index_en.htm, 2004.

[RSP 04b] RADIO SPECTRUM POLICY GROUP, "Report on Study on conditions and options in introducing secondary trading of radio spectrum in the European Community", European Community, July 2004, http://europa.eu.int/information_society/policy/radio_spectrum/highlights/what_new/index_en.htm.

[SDR] SDR Forum, http://www.sdrforum.org.

[SPE 03] SPECTRUM TRADING WORKSHOP, Econ, Analysis, HH, Brussels, Belgium, 11 December 2003.

[STA 04] STAPLE G., WERBACH K., "The End of Spectrum Scarcity", *IEEE Spectrum Magazine*, p. 41-44, March 2004.

[VAL 01] VALLETTI T.M., "Spectrum trading", *Telecommunications Policy*, vol. 25, no. 10-11, p. 655-670, 2001.

[VAN 98] DE VANY A., "Implementing a Market Based Spectrum Policy", *Journal of Law and Economics*, 41(2), p. 627-646, 1998.

[WEB 98] WEBB W., "The role of economic techniques in spectrum management", *IEEE Comm. Magazine*, 36, p. 102-107, March 1998.

[WEI 04] WEISS T.A., JONDRAL F.K., "Spectrum pooling: an innovative strategy for the enhancement of spectrum efficiency", *IEEE Communications Magazine*, vol. 42, no. 3, p. 8-14, March 2004.

[WER 02] WERBACH K., "Open Spectrum – The New Wireless Paradigm", New America Foundation, Spectrum Policy Program, Spectrum Series Working Paper, no. 6, October 2002.

[WRC 03] WRC, Resolution 951 [COM7/2], WRC-03, Options to improve the international spectrum regulatory framework, Geneva, 2003.

[XG] XG, American defense project DARPA XG, www.darpa.mil/ato/programs/XG.

[ZAN 97] ZANDER J., "Radio Resource Management in Future Wireless Networks", *IEEE Communications Magazine*, August 1997.

[ZEK 03] ZEKAVAT S.A., NASSAR C.R., "Spectral sharing in multi-system environments *via* multi-carrier CDMA", *IEEE ICC'03*, vol. 3, p. 2223-2228, 11-15 May 2003.

List of Authors

Marylin ARNDT
France Télécom R&D
Meylan
France

Eric BATUT
France Télécom R&D
Meylan
France

Christian BONNET
Eurecom Institute
Sophia Antipolis
France

Hervé CALLEWAERT
Eurecom Institute
Sophia Antipolis
France

Jean-Yves CHOUINARD
Laval University
Quebec
Canada

Jean-Luc DANGER
ENST
Paris
France

Antoine DELAUTRE
Thales Communications
Colombes
France

Yann DENEF
Thales Communications
Colombes
France

Guillaume DORBES
Alcatel
Marcoussis
France

Jean-Philippe FASSINO
France Télécom R&D
Meylan
France

Lionel GAUTHIER
Eurecom Institute
Sophia Antipolis
France

Florence GERMAIN
France Télécom R&D
Meylan
France

David GRANDBLAISE
Motorola
Gif-sur-Yvette
France

Tahar JARBOUI
France Télécom R&D
Meylan
France

Raymond KNOPP
Eurecom Institute
Sophia Antipolis
France

Ioannis KRIKIDIS
ENST
Paris
France

Marc LACOSTE
France Télécom R&D
Meylan
France

Christian LEREAU
France Télécom R&D
Meylan
France

Patrick LOUMEAU
ENST
Paris
France

François MARX
France Télécom R&D
Meylan
France

Pascal MAYANI
Eurecom Institute
Sophia Antipolis
France

Aawatif MENOUNI HAYAR
Eurecom Institute
Sophia Antipolis
France

Benoît MISCOPEIN
France Télécom R&D
Meylan
France

Jean-Francois NAVINER
ENST
Paris
France

Lírida NAVINIER
ENST
Paris
France

Dominique NUSSBAUM
Eurecom Institute
Sophia Antipolis
France

Jacques PULOU
France Télécom R&D
Meylan
France

Sébastien ROY
Laval University
Quebec
Canada

Guillaume VIVIER
Motorola
Gif-sur-Yvette
France

Michelle WETTERWALD
Eurecom Institute
Sophia Antipolis
France

Index

A

B

C

D

E

I

J

M

P

R

S

U

V

X